Technological Innovations and Advances in Hydropower Engineering

Edited by Yizi Shang, Ling Shang and Xiaofei Li

Published in London, United Kingdom

IntechOpen

Supporting open minds since 2005

Technological Innovations and Advances in Hydropower Engineering
http://dx.doi.org/10.5772/intechopen.94652
Edited by Yizi Shang, Ling Shang and Xiaofei Li

Contributors
Emanuele Quaranta, Pavel N. Terskii, Olga V. Gorelits, Galina S. Ermakova, George Mamulashvili, Hemlal Bhattarai, Ling Shang, Haifeng Shi, Feng Kong, Ying Wang, Yizi Shang, Xiaofei Li

First published in London, United Kingdom, 2022 by IntechOpen
IntechOpen is the global imprint of INTECHOPEN LIMITED, registered in England and Wales, registration number: 11086078, 5 Princes Gate Court, London, SW7 2QJ, United Kingdom
Printed in Croatia

British Library Cataloguing-in-Publication Data
A catalogue record for this book is available from the British Library

Additional hard and PDF copies can be obtained from orders@intechopen.com

Technological Innovations and Advances in Hydropower Engineering
Edited by Yizi Shang, Ling Shang and Xiaofei Li
p. cm.
Print ISBN 978-1-83968-914-7
Online ISBN 978-1-83968-915-4
eBook (PDF) ISBN 978-1-83968-916-1

For EU product safety concerns:
IN TECH d.o.o., Prolaz Marije Krucifikse Kozulić 3, 51000 Rijeka, Croatia, info@intechopen.com or visit our website at intechopen.com.

Meet the editors

Prof. Dr. Yizi Shang is a pioneering researcher in hydrology and water resources who has devoted his research career to promoting the conservation and protection of water resources for sustainable development. He is presently associate editor of *Water International* (official journal of the International Water Resources Association). He was also invited to serve as an associate editor for special issues of the *Journal of the American Water Resources Association*. He has served as an editorial member for international journals such as *Hydrology*, *Journal of Ecology & Natural Resources*, and *Hydro Science & Marine Engineering*, among others. He has chaired or acted as a technical committee member for twenty-five international forums (conferences). Dr. Shang graduated from Tsinghua University, China, in 2010 with a Ph.D. in Engineering. Prior to that, he worked as a research fellow at Harvard University from 2008 to 2009. Dr. Shang serves as a senior research engineer at the China Institute of Water Resources and Hydropower Research (IWHR) and was awarded as a distinguished researcher at National Taiwan University in 2017.

Prof. Dr. Ling Shang has carried out extensive field and model studies on artificial intelligence, big data, and edge computing related to the intelligent operation of various engineering projects. His most recent work focuses on the joint operation of cascaded hydropower projects and their impact on efficiency improvement. He was elected a distinguished member of the China Computer Federation (CCF) and appointed the head of Key Laboratory of Artificial Intelligence Research, Nanjing Vocational College of Information Technology, China. He received a BEng from Zhengzhou University, China, in 2003, MEng from Hohai University, China, in 2006, and Ph.D. in Engineering from Université de Lille, France, in 2011.

Prof. Li Xiaofei is the director of the Architectural Research Institute, SANY Construction Technology Co., Ltd., China. Since 2009 she has been involved in the construction of many building projects across China, including hydropower projects in Yellow River, Yangtze River, and Haihe River basins. She has led the team to overcome difficulties in the electrical engineering planning and management field and made large-scale hydropower projects optimally operated. In 2020, she obtained a Registered Electrical Engineer certificate. Prof. Xiaofei holds four registered patents and has published eleven papers.

Contents

Preface

The hydropower sector has gone through more than 140 years since the world's first hydropower station was established in 1878. Hydropower construction in most developed countries unfolded from the 1920s to the 1960s and entered a stable development stage in the 1970s, with hydropower resources in Switzerland and France almost fully exploited in the 1980s. The upsurges of hydropower construction in Asia, Africa, Latin America, and the United States began in the 1960s. Since then, emerging and developing economies have been leading global hydropower growth, while the hydropower infrastructure of developed countries is gradually aging. The average service life of hydropower stations is close to 50 years in North America and 45 years in Europe. Signs of risks from aging infrastructure are observed all over the world. Examples include the flooding of the Sayano-Shushenskaya Hydropower Station in Russia in August 2009 and the damage to the spillways of the Oroville Dam in the United States in 2017, prompting the evacuation of residents around the dam.

Hydropower has represented a decreasing share of power generation in advanced economies since 2000. Even so, global installed hydropower capacity has increased by 70% over the past two decades. Globally, about half of the economic potential of hydropower remains untapped, with a particular high of nearly 60% in emerging and developing economies. In addition, there are already a massive number of hydropower facilities that have been providing affordable and reliable renewable power on demand for decades. Modern upgrades are needed to ensure that they can contribute to power security in a sustainable manner in the coming decades.

Hydropower is not only a maturely used clean energy source, but it is also a highly flexible energy storage system. Compared with nuclear power, coal power, and even gas-fired power, hydropower is quicker in regulating electricity production, so it can efficiently serve peak shaving in the future when wind power, solar, and other intermittent power sources are applied on a large scale. In 2020, hydropower accounted for 17% of global power generation, the third-largest source of electricity after coal and natural gas. From 2021 to 2030, global installed hydropower capacity is expected to keep expanding by 17% to reach 230 GW, according to a report from the International Energy Agency. In this context, a challenge for global sustainable hydropower development will be to support the healthy and rapid growth of hydropower in emerging and developing economies and assist developed countries in implementing relatively robust upgrades to hydropower facilities.

This book introduces technological innovations in hydropower engineering and their contributions to rapid and sustainable hydropower development. It consists of six chapters, that cover the leapfrog development of hydropower in China, hydropower station operation and intelligent management, research on new methods of hydropower utilization, research on the optimized operation of hydropower stations, case studies of hydropower technology innovations, and sustainability of hydropower generation.

China is undoubtedly the fastest, though not the earliest, in hydropower development in the world. As of the end of 2020, China's installed hydropower capacity reached 370 GW, ranking top in the world for sixteen consecutive years. Chapter 1 "Hydropower Development in China: A Leapfrog Development Secured by Technological Progress of Dam Construction" reviews hydropower development in China, specifically its post-1970s super hydropower projects, and elaborates on the key support of technological progress in dam construction to hydropower projects. In addition, this chapter explains the problems brought about by rapid hydropower development in China and gives in-depth insights and explorations into the country's future hydropower development. These valuable successful experiences can enlighten other countries as to sustainable hydropower development and utilization.

Russia's Volga-Kama Cascade has been in operation for fifty years. In recent years, changes in the global climate and external environment have not only reduced the efficiency of cascade power generation but have also led to a drop in the guaranteed rate of reservoir water supply. In addition, frequent extreme events also threaten the cascade safety and cause ecological disasters. Chapter 2 "Hydropower in Russia: Case Study on Hydrological Management of the Volga-Kama Cascade," using the Volga-Kama Cascade as an example, warns that old hydropower stations should adaptively transform infrastructure in a changing environment and tries to solve the safety problems and rejuvenate old hydropower stations through the adjustment of operation and management methods. This chapter provides an important reference for the study of old hydropower stations in other countries in the world.

Current hydropower generation methods worldwide generally convert the gravitational potential energy of water into electrical energy, but this should not be the whole of hydropower generation. Chapter 3 "Hydro Power Tower (HYPOT)" proposes a technical method of generating electricity from the horizontal flow of water, which can convert the horizontal kinetic energy of seawater into electrical energy. This method inspires us to tap the undiscovered energy of water resources and explore new ways to harness water energy.

During the process of hydropower generation, hydropower stations are subject to changing external environmental factors, including reservoir inflow changes and electricity load fluctuations. Improving the efficiency of hydropower generation under the influence of many uncertain factors has been a persistent challenge to efficient hydropower development and utilization. Chapter 4 "Improved Memetic Algorithm for Economic Load Dispatch in a Large Hydropower Plant" introduces intelligent algorithms into the operation and management of China's Three Gorges Hydropower Station. Practice or experiment shows that the improved memetic algorithm (IMA) can indeed raise the power generation efficiency of hydropower stations. This offers a new technical solution that optimizes the dispatch and operation of large hydropower stations by using intelligent algorithms.

Hydropower development should not compromise ecological environment. Chapter 5 "Innovative Projects and Technology Implementation in the Hydropower Sector" analyzes the eco-environmental impacts of hydropower development and hydropower station construction. It also introduces some technical methods for improving power generation equipment for the purpose of eco-environmental protection, noting that eco-environmental protection is an important issue urgently to be addressed in future hydropower development. These innovations not only protect the species diversity of river ecosystems but also ensure the realization of the expected economic and social benefits of hydropower stations. They embody the significant research on the application of environmental protection technology in hydropower generation.

Hydropower is the world's largest clean energy that has realized commercial development on a large scale, making an important contribution to cutting carbon emissions. Chapter 6 "Hydropower and Sustainability" examines the promotion and restraint of global climate change, efficient hydropower development and utilization, and energy structure adjustment. It also highlights that the concept of sustainability should be implemented in all stages of planning, construction, and operation. It is gratifying that sustainability has gradually become a global consensus.

This book shares the latest progress in scientific research on hydropower and uses some practical cases to inspire innovative ideas for future hydropower research. This book received support from the Beijing Natural Science Foundation (Grant number: JQ21029).

Dr. Yizi Shang
Professor,
China Institute of Water Resources and Hydropower Research,
Beijing, China

Dr. Ling Shang
Professor,
Faculty of Network and Communication,
Nanjing Vocational College of Information Technology,
Nanjing, China

Xiaofei Li
Professor,
Architectural Research Institute,
SANY Construction Technology Co., Ltd,
Beijing, China

Section 1

Hydropower Construction and Renewal

Chapter 1

Hydropower Development in China: A Leapfrog Development Secured by Technological Progress of Dam Construction

Yizi Shang, Xiaofei Li and Ling Shang

Abstract

It has been over 110 years since China's first hydropower station, Shilongba Hydropower Station, was built in 1910. With the support of advanced dam construction technology, the Chinese installed capacity keeps rising rapid growth, hitting around 356 GW nationwide by the end of 2019, and the annual electricity production exceeds 10,000 TWh. At present, China contributes to 25% of global installed hydropower capacity, ranking first in the world for 20 consecutive years since 2001 and surpassing the combined of the 4 countries ranking second to fifth. This paper reviews China's progress in the context of global hydropower development and examines the role of technological advance in supporting China's hydropower projects, especially dam construction technology. China is currently actively promoting the "integration of wind, solar, hydro, and coal power generation and energy storage" and building a smart grid of multi-energy complementary power generation. New technologies and new concepts are expected to continue to lead the world's hydropower development trends.

Keywords: China, hydropower, super hydropower project, installed hydropower capacity, dam construction technology, high dam and large reservoir

1. Introduction

Hydropower is a clean and renewable energy source among conventional energy sources and has the advantages of low operating cost, simple electromechanical equipment, and operational flexibility [1–5]. Hydropower development has thus emerged as a priority option in most developed countries [6–8]. The utilization rate of hydropower resources in developed countries such as France and Switzerland already hit 97% in the late 1980s. In developing countries, the process of water resource development was slow in the past with a low degree of hydropower development due to political, economic, and other reasons. There has been an evidently rapid increase in the pace and rate of hydropower development and utilization in the recent four decades, especially since the mid-1980s. By the end of 2020, the total installed hydropower capacity in the world has reached 1330 GW, and the installed hydropower capacity in 2020 will increase by 1.6%. The global hydropower generation accounts for 16% of the total global power generation, which is lower than coal-fired power generation and gas-fired power generation, ranking third in the world. Among them, East Asia and the Pacific have the most hydropower generation, accounting for 37.6% of the global total. Major contributors to the added installed capacity are China, Laos, Pakistan in Asia, Brazil in South America, Angola, Uganda, and Ethiopia in Africa, and Turkey in Europe.

Dam is the principal structure for hydropower generation, so the number of dams speaks for the activity level of hydropower development in a country. A large dam is defined by the International Commission on Large Dams (ICOLD) as any dam above 15 m in height (measured from the lowest point of foundation to top of dam) or any dam between 10 m and 15 m in height impounding more than 3 million m^3. According to this definition, there were 58,713 large dams worldwide in 2020. China embraces 23,841 large dams, the most among all countries and more than the combined of the following United States (9263), India (4407), and Japan (3130) [9]. It ranks the world's first with a share of more than 40%. In fact, in addition to storing water for power generation, dams can alter natural runoff through reservoir regulation to render functions such as water supply from upstream reservoirs, downstream navigation, and river ecological flow maintenance. This helps alleviate the plague associated with flood, drought, electricity shortage, and water environmental degradation. In particular, hydropower development avoids greenhouse gas (GHG) emissions of thermal power generation. In view of this, the United States has incorporated hydropower in many targets of the 17 Sustainable Development Goals (SDGs) published in September 2015 [10]. More than 100 countries have so far made it clear that they will continue to build dams and vigorous develop hydropower. It is estimated that by 2035, global installed hydropower capacity will add by about 480 GW to reach 1750 GW, with annual electricity generation of 6100 TWh and hydropower exploitation rate of 38.6%; and by 2050, global installed hydropower capacity will further grow by 300 GW to reach 2050 GW.

China has set an example for global hydropower development. At the end of 2019, China's installed hydropower capacity hit 356 GW with an electricity output of 1300 TWh, accounting for 27.2% and 30.8% of the global total, respectively (**Table 1**). Such a scale is more than three times that of the United States, and larger than the combined of countries ranking second to fifth. China has grown into a veritable hydroelectricity powerhouse based on many super large dams and reservoirs. Next, China's course and

Year	China's hydropower generation TWh	China		World	
		Electricity generation/TWh	%	Hydropower production/TWh	%
2020	1322	7779.1	17.0	4296.8	30.8
2019	1269.7	7503.4	16.8	4222.2	30.1
2018	1202.4	7111.8	16.9	4193.1	28.7
2017	1155.8	6495.1	17.8	4059.9	28.5
2016	1051.8	5911.1	17.8	4138.4	25.4
2015	1180.7	5810.5	20.3	4032.1	29.3
2014	1111.7	5649.6	19.7	3970.5	28.0
2013	920.3	5397.6	17.1	3847.0	23.9
2012	855.6	4986.5	17.2	3765.3	22.7
2011	668.1	4730.6	14.1	3407.2	19.6

Table 1.
Share of hydropower in China's electricity generation and global hydropower generation.

achievements in hydropower development will be reviewed, and potential challenges to China's sustainable hydropower development will be analyzed [11].

2. China's course of hydropower development

It has been completely 110 years since China's first hydropower station, Shilongba Hydropower Station, was constructed in 1910 [12–16]. At of the end of 2019, there were 46,758 hydropower stations nationwide, with a total installed capacity of 332.89 GW. Among them, 22,190 above the designated size provide 327.3 GW and 24,568 below the designated size provide 5.59 GW. Besides, 11 of the world's top 20 hydropower stations are located in China (**Table 2**), and all the super hydropower stations built after 1990 come from China without exception.

World ranking	Name	Country	Installed capacity	Annual production	Date of start of operation	River
			10 MW	100 GWH		
1	Three Gorges Dam	China	2250	847	2003	Yangtze River
2	Itaipu	Brazil/ Paraguay	1400	900	1983	Parana River
3	Xiluodu	China	1386	571	2014	Jinsha River
4	Baihetan	China	1250	640	2018	Jinsha River
5	Wudongde	China	1020	387	2020	Jinsha River
6	Guri	Venezuela	910	510	1968	Rio Caroni
7	Tucurui	Brazil	837	324	1984	Tocantins River
8	Xiangjiaba	China	775	307	2012	Jinsha River
9	La GrandeII	Canada	732	358	1979	La Grande
10	Grand Coulee	America	649	203	1942	Columbia River
11	Sayano-Shushenskaya	Russia	640	235	1978	Yenisei
12	Longtan Dam	China	630	187	2007	Hongshui River
13	Krasnoyarsk	Russia	600	204	1967	Yenisei
14	Nuozhadu	China	585	239	2012	Lancang River
15	Churchill Falls	Canada	542	345	1972	Churchill River
16	Jinping-II	China	480	242	1967	Yalong River
17	Bratsk	Russia	450	226	1967	Angara River
18	Xiaowan Dam	China	420	185	2009	Lancang River
19	Laxiwa	China	420	102	2009	Yellow River
20	Ertan Dam	China	330	170	1998	Yalong River

Table 2.
World's top 20 hydropower stations.

Those Chinese hydropower stations are at the forefront of the world in terms of technology, representing important milestones in China and even the World's hydropower development. **Figure 1** shows the time points of project construction for China's super hydropower stations and their locations in China are as shown in **Figure 2**.

Shilongba Hydropower Station, the first hydropower station in China, was constructed in 1910 and commissioned in 1912. Its installed capacity was 480 kW upon completion and rose to 360 MW in 1949 when New China was founded. Sanmenxia Hydropower Station, the first large-scale hydropower station built in New China, started construction in April 1957 and operation in April 1961, with an installed capacity of 1160 MW. The multi-year average annual output stands at 6 TWh. It contains a concrete gravity dam with a maximum height of 106 m.

Xin'anjiang Hydropower Station was built at the same time as Sanmenxia Hydropower Station and officially put into operation 1 year earlier. This concrete gravity dam, 105 m tall with a crest length of 466.5 m, enables an installed capacity of 662.5 MW, based on 40,000 tons of metal structures and electromechanical equipment, after 5.8592 million m^3 of earth was moved and 1.755 million m^3 of

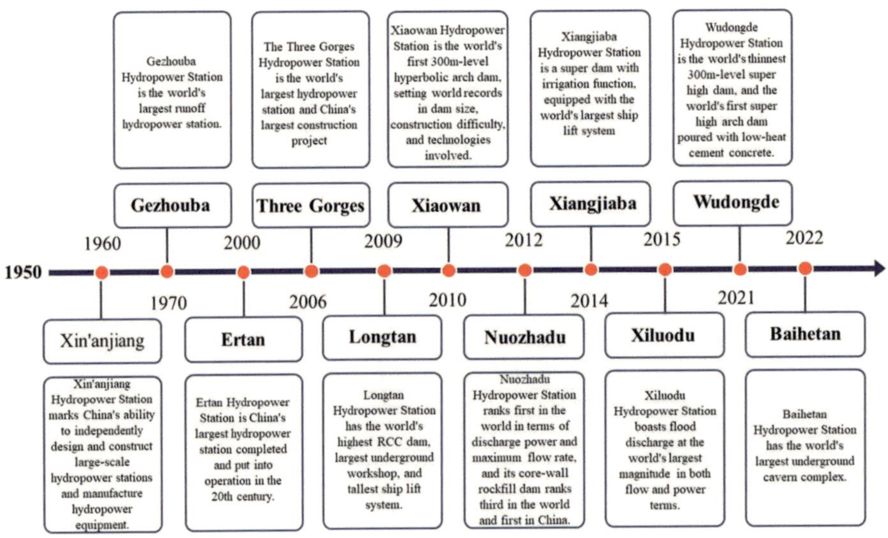

Figure 1.
Time points of project construction for China's super hydropower stations.

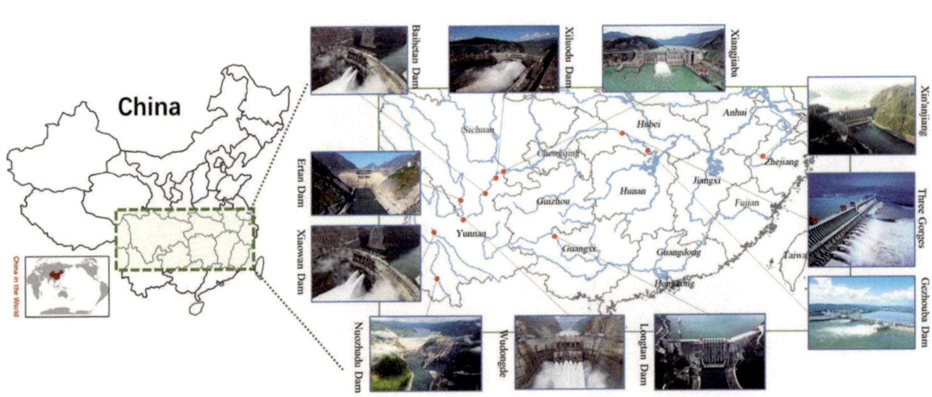

Figure 2.
Location map of China's super hydropower projects stations.

concrete poured. Xin'anjiang is a milestone in China's hydropower development that marks China has become able to independently design and construct large-scale hydropower stations and manufacture hydropower equipment.

Gezhouba Hydropower Station is the world's largest runoff hydropower station. The construction of the Gezhouba Water Control Project spanned from December 1970 to late 1988, with Phase I completed in 1981 and Phase II starting in 1982. The project consists of ship locks, power stations, spillway sluices, scouring sluices, and auxiliary dams. The dam is a gate dam with a maximum height of 47 m. Two river-bed power stations are located in Erjiang and Dajiang. The former has an installed capacity of 965 MW sourced from two sets of 170 MW generating units and five sets of 125 MW generating units. The latter is equipped with 14 sets of 125-MW hydropower generator units to form an installed capacity of 1750 MW. They make Gezhouba's installed capacity total 2715 MW. The 170 MW generating units of the Erjiang Power Station have a turbine diameter of 11.3 m and a stator outer diameter of 17.6 m.

Ertan Hydropower Station is China's largest power station built and commissioned in the twentieth century. The construction began in September 1991 and ended in 2000, and the first generating unit started operation in July 1998. Ertan Hydropower Station is located at the junction of Yanbian and Miyi counties, Panzhihua City, southwest border of Sichuan Province, China. Ertan Dam, with a maximum height of 240 m, sits on the lower Yalong River, 33 km away from the intersection of the Yalong River and the Jinsha River and 46 km from Panzhihua City. Ertan is the first of cascade hydropower stations developed in the Yalong River Hydropower Base, with Guandi in the upstream and Tongzilin in the downstream. Given a normal pool level of 1200 m above sea level, the reservoir impounds 5.8 km^3 of water, including a regulated storage capacity of 3.37 km^3. The installed capacity totals 3.3 GW with a guaranteed output of 1 GW, and the annual electricity production averages 17 TWh. The project involving 28.6 billion yuan in investment renders comprehensive benefits in addition to power generation.

The Three Gorges Hydropower Station is the world's largest hydropower station and China's largest construction project. It secures an installed capacity of 22.4 GW with an average annual output of 90 TWh by installing 32 sets of 700-MW generating units. The construction started in 1994 and was officially finished in 2006 [17]. The concrete gravity dam, the world's largest of its kind, is 2335 m long, 115 m wide at the bottom and 40 m wide at the top, and 185 m above sea level, with normal storage level of 175 m. It can withstand floods so severe they come only once in 10,000 years owning to the designed maximum outflow of 100,000 m^3 per second. The whole project moved about 134 million m^3 of earth and stone and used about 28 million m^3 of concrete and 593,000 tons of steel. The reservoir is over 600 km in length and 1.1 km in width on average, making a surface area of 1084 km^2. It impounds 39.3 billion m^3 of water, including 22.15 billion m^3 for flood control in a seasonal manner. Totally, 32 sets of 700-MW generating units are deployed on the back of the dam, with 14 sets on the left bank, 12 sets on the right bank, and 6 sets underground, and in addition two sets of 50-kW power supply units form an installed capacity of 22.5 GW, ranking second in the world and far exceeding that of Brazil's Itaipu Hydropower Station. By 24 o'clock, December 31, 2014, the Three Gorges Hydropower Station generated 98.8 TWh of electricity throughout the year, a new world record that secures its first position in terms of annual output. This is equivalent to a reduction of nearly 100 million tons of carbon dioxide (CO_2) emissions from over 49 million tons of raw coal consumption.

Longtan Hydropower Station started construction on July 1, 2001 and was completed and commissioned at the end of 2009. With a designed storage level of 400 m, the 216.5 m high and 836 m long dam has a storage capacity of 27.3 billion m^3, an installed capacity of 6.3 GW, and an annual output of 18.7 TWh. This dam sets three new world records: the highest roller-compacted concrete (RCC) dam (with a maximum height of 216.5 m, a crest length of 832 m, and a concrete volume of 7.36 million m^3); the largest underground workshop (385 m long, 28.5 m wide, and 74.4 m high); and the tallest ship lift system (with a full length of over 1800 m and a maximum lifting height of 156 m by two steps).

Xiaowan Hydropower Station is built primarily for power generation but also performs functions in flood control, irrigation, and water transportation. With a maximum height of 294.5 m, the world's tallest arch dam also ranks first in key indicators of arch dam construction such as peak ground acceleration, crest length, and water thrust. Construction started on January 1, 2002. River closure was achieved on October 25, 2004, a year ahead of schedule, and concrete pouring for the first dam warehouse began on December 12, 2005. The diversion tunnel was closed for water storage on December 16, 2008. Pouring and capping across the board was completed on March 8, 2010, marking the birth of the world's tallest 300-m-level hyperbolic arch dam. All the six generating units with a combined capacity of 4.2 GW were commissioned on August 22, 2010. Xiaowan Reservoir, the first reservoir of cascade power stations, impounds about 15 billion m^3 of water, including nearly 10 billion m^3 for multi-year regulation. The power station is equipped with six mixed-flow generators with a unit capacity of 700 MW and thus forms a total installed capacity of 4200 MW with a guaranteed output of 1854 MW. The average annual electricity production reaches 19.06 TWh.

Xiangjiaba Hydropower Station is a super dam with irrigation function, equipped with the world's largest ship lift system. It is only 1500 m away from Shuifu City, located on the lower Jinsha River at the junction of Shuifu City, Yunnan Province and Xuzhou District, Yibin City, Sichuan Province. The construction was formally kicked off in November 2006, and the station was fully put into operation in July 2014. The installed capacity reaches 7.75 GW with an average annual output of 30.7 TWh, including eight sets of 800-MW rectangular turbines and three sets of 450-MW large turbines.

The core-wall rockfill dam of Nuozhadu Hydropower Station is a classic case of China's successful application of gravelly clay core wall to effectively improve the strength of earth-rock dams. It is of great significance as the first successful case. The construction started in January 2006 and ended after final acceptance in May 2016, with the first generating units put into operation in August 2012. The normal storage level of the reservoir is 812 m and the maximum height of the dam is 261.5 m, ranking third in the world and first in China among similar dams. The open spillway is 1445 m long and 151.5 m wide, the largest in Asia. Flood discharge can reach a magnitude of 55.86 GW with a maximum flow rate of 52 m per second, ranking first in the world in both terms. The stratified water intake scheme adopted by Nuozhadu Hydropower Station created a precedent for environmentally friendly design of hydropower generation in China. The project involving an investment of about 61.1 billion yuan realizes an average annual electricity production of 23.912 TWh based on 4088 utilization hours, which is equivalent to saving 9.56 million tons of coal equivalent and reducing CO_2 emissions by 18.77 million tons each year.

Construction of Wudongde Hydropower Station was kicked off in 2015. Concrete was poured for dam construction in 2017. The first generating units were commissioned in June 2020, and all the generating units were officially put into operation in June 2021. A concrete hyperbolic arch dam with a crest elevation of 988 m, a maximum height of 270 m, and a base thickness of 51 m is used as the water-retaining structure. The thickness to height ratio of only 0.19 makes it the world's thinnest 300-m-level super high dam. Wudongde Dam is also the world's first super high arch dam poured with low-heat cement concrete. Twelve generating units with a unit capacity of 850 MW have been installed in the power station, making a total of 1.02 GW, the fourth largest in China and the seventh largest in the world.

Xiluodu Hydropower Station focuses on power generation but also contributes to flood control, sand retention, improvement of upstream shipping conditions, and cascade compensation for downstream power stations. It is located on the Jinsha River at the junction of Sichuan and Yunnan. The construction started in June 2004 and fully ended in 2015. A total of 18 sets of 770-MW generating units have been installed, forming an installed capacity of 13.86 GW that supports an average annual output of 57.1 TWh. Flood discharge is a major highlight of Xiluodu Dam, with the flow and power of discharge far exceeding the highest level of arch dams in the world. In European and American countries that have led the world in dam construction, arch dams generally do not have drainage holes out of consideration of stability, such as the famous Hoover Dam. In contrast, a large number of holes are opened in the Xiluodu Dam when the spillway tunnels on both sides of the dam are not enough to discharge all the floods. The Xiluodu Hydropower Station project has the characteristics of narrow river valley, high arch dam, huge volume of discharge, multiple generating units, large cavern complex, and high seismic resistance capacity. It has outperformed existing projects in many key technologies, boasting world's highest level of comprehensive technical difficulty.

Baihetan Hydropower Station is designed primarily for power generation but also plays a role flood control, sand retention, improvement of downstream shipping conditions, and development of navigation in the reservoir area. Equipped with 16 sets of 1-GW Francis turbine generators, Baihetan ranks second in the world in terms of total installed capacity and first in terms of unit installed capacity. The project was officially kicked off in 2013 and is expected to be completed in 2022, with the first group of generating units formally put into operation in July 2021. The underground cavern complex has a total length of 217 km, the largest in the world. For the first time, the power station uses all Chinese-made GW-level turbine generators. This is another historic leap for China's major hydropower equipment after the localization of generating units in the Three Gorges Hydropower Station and 800-MW generating units in Xiangjiaba Hydropower Station.

3. Super technologies underpinning super projects

Taking into account the historical background of hydropower development, China's super hydropower projects are underpinned by its unique super dam construction technologies, as shown in **Figure 3**.

Since the founding of New China in 1949, substantial progress has been made in the construction of gravity dams [18–21], arch dams [21–24], and earth-rock dams [25–27].

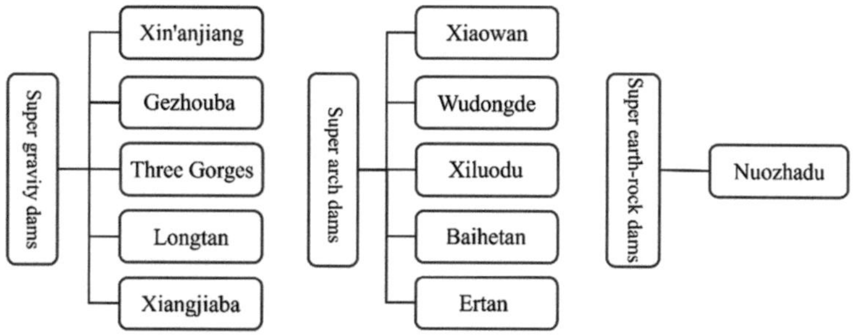

Figure 3.
China's super hydropower stations by world indicators.

Such advancement of dam construction technologies provides the basic guarantee for China's super hydropower projects.

3.1 Super gravity dams

Concrete gravity dams with simple structure and clear stress can integrate various spillway combinations to ensure high reliable resistance to flood hazards [28, 29]. They are well adapted to terrain and geological conditions by arranging flexibly various types of power plants. In the early 1930s, the 221-m high Hoover Dam was built on the Colorado River in the United States, marking the arrival of a rapid development period of dam construction. In the 1980s, a group of ultra-high gravity dams (higher than 200 m) embodying the dam construction technology of the twentieth century was commissioned to play an important role in river flow regulation, flood control, power generation, irrigation, and water supply. The 285-m high Grand Dixence Dam built in Switzerland in the early 1960s remains to be the world's tallest concrete gravity dam.

In China, the construction of gravity dams in the modern sense began after the founding of New China. In the 1950s, two slotted gravity dams, namely Xin'anjiang and Gutian-I, were constructed to meet the needs of economic and social development. In the 1960s, another two slotted gravity dams, i.e. Yunfeng and Danjiangkou, and two solid gravity dams, i.e. Liujiaxia and Sanmenxia, were added. Hunanzhen trapezoidal gravity dam was completed in the 1970s and Wujiangdu arch gravity dam in the early 1980s. There were over 20 gravity dams taller than 70 m in the country by the 1980s. In this stage, efforts were made to explore ways to reduce dam engineering volume and save project investment, which makes possible the design and construction of many new and lightweight gravity dams.

Dam construction technology, including concrete gravity dams, made notable progress after the 1980s. Following the gravity dam projects in Shuikou, Geheyan, Wuqiangxi, and Yantan, a number of high gravity dam projects such as the Three Gorges Dam, Longtan, Guangzhao, Jinanqiao, and Xiangjiaba were successively launched at the turn of the century, constantly setting records in terms of dam height and project scale. Longtan Dam on the Hongshui River is currently the highest gravity dam in China, with a designed height of 216.5 m and a concrete volume of 7.5 million m^3. In this stage, the height of concrete gravity dams in China grew from 100 m to 150 m and further to 200 m. The number of solid gravity dams also increased due to construction efficiency improvement because such simple structures are more suitable for mechanized construction. A series of new energy dissipation works were commissioned, solving the problem of high-head and large-flow flood discharge and energy dissipation. In addition, RCC gravity dams tend to gradually replace normal concrete gravity dams. China has technologically reached the international advanced level in the construction of high concrete gravity dams and high RCC gravity dams [30, 31]. The most representative examples are undoubtedly the Three Gorges Dam (normal concrete gravity dam) and Longtan Dam (RCC gravity dam).

The Three Gorges Dam is the most concrete incorporated gravity dam in the world. The dam with a height of 181 m and a crest length of 2309.47 m uses 16 million m^3 of concrete among the project total of 28 million m^3. Two famous dams comparable to the Three Gorges Dam are the 168 m high Great Coulee Dam on the Columbia River in the United States (with a concrete volume of 7.26 million m^3) and the 162 m high Guri Dam on the Caroni River in Venezuela (with a concrete volume of 6.71 million m^3). Longtan Dam as RCC gravity dam represents the highest level of RCC dam construction in the world. Built on the Hongshui River, the dam has a designed height of 216.5 m and a concrete volume of 7.5 million m^3. In order to solve the constraints of high temperature and rain and shorten the time of dam construction, China developed the world's most advanced technology of RCC construction under special climates. This technology effectively controls the initial setting time of concrete and realizes the quick coverage by improving concrete production and transportation capacity and silo capacity. It is integrated with unique interlayer treatment technology to enable construction in high temperature and rainy conditions. In addition, the pioneering anti-seepage scheme that combines second-grade RCC and distorted concrete also represents the world's highest level.

3.2 Super arch dams

Arch dams are water-retaining structures curving upstream on the plane, where water thrust is transmitted partially or fully through the arch to the bedrock on both sides of the river valley. The stability of arch dams is largely supported by the reaction of the bedrock of arch side to water pressure, rather than dam weight as in the case of gravity dams. Axial reaction force at the section of arch ring can take advantage of the strength of dam materials. Therefore, arch dams perform well in terms of economy and safety [32–34].

In the 1980s, RCC began to be applied to arch dams. In 1988, the world's first RCC arch dam, Knellpoort, was built in South Africa. In 1993, China completed its RCC gravity arch dam called Puding by using new dam construction technologies. Thereafter, a group of hydropower stations such as Longyangxia, Ertan, Xiaowan, Laxiwa, Jinping-I, and Xiluodu have been gradually put in place. Baihetan Hydropower Station is still under construction, expected to be completed in 2022. Longyangxia Hydropower Station is the first large-scale cascade power station on the upper Yellow River. It consists of barrage dam, waterproof structures, and powerhouse. The dam is 178 m high, 1226 m long (including the 396-m-long main dam), and 23 m wide. The project can not only block all the annual flow of 130,000 m^3 from the upper Yellow River but also forms an reservoir with a surface area of 383 m^2 and a storage capacity of 24.7 billion m^3.

Ertan Hydropower Station is the largest hydropower station built and commissioned in China in the twentieth century. The 240-m high dam, whose construction took 10 years, was the tallest arch dam in Asia at that time. The double-curvature arch dam, China's tallest dam, ranks third among dams of its kind in the world, topping in terms of crest length and flood discharge capacity. Laxiwa Hydropower Station is the largest hydropower station and clean energy base on the Yellow River. The dam is 250 m high, but only 49 m wide at the bottom. It is deemed as a thin dam as the ratio of width to height is 0.196, lower than the national standard 0.2. Nevertheless, it renders the largest installed capacity and electricity production in the Yellow River Basin, in addition to the highest dam. Xiaowan Dam with a maximum height of 294.5 m is the world's highest arch dam under construction. It outperforms in this category worldwide in key indicators such as peak ground acceleration, crest length, and water thrust.

Jinping-I Hydropower Station contains a double-curvature arch dam with a height of 305 m, the tallest of its kind in the world. Wudongde Hydropower Station features a super high arch dam, the world's thinnest at the 300 m level and the world's first poured with low-heat cement concrete throughout the whole dam.

The concrete double-curvature arch dam at Xiluodu Hydropower Station has a maximum height of 278 m, a crest elevation of 610 m, and an arc length of 698.07 m, which enables it to withstand 15 million tons of water thrust. The arch dam of Baihetan Hydropower Station has a crest elevation of 834 m and a maximum height of 289 m. The arch dam of Laxiwa Hydropower Station is 250 m high at the most.

3.3 Super earth-rock dams

Earth-rock dams refer in general to water-retaining structures built by dumping and compacting locally available earth, rock, or mixture. Earth dams are made up mostly of earth and gravel, and rockfill dams are made up mostly of gravel, pebbles, and crushed stones. Earth-rock dams as the oldest type of dams contain both two kinds of materials. Modern technology for earth-rock dams has been developed since the 1950s, which enables the construction of several high dams. Earth-rock dams are now among the most widely used and fastest-growing dam types in the world owning to strong adaptability to complex geological conditions, local availability of materials, and small investment [26, 35].

The United States, Canada, and the former Soviet Union has made rapid progress in earth-rock dams since early twentieth century. A number of 200 m–300 m-level high dams have been built, as Oroville (235 m high) in the United States, Boruca (267 m high) in Costa Rica, and Nurek (300 m high) in the former Soviet Union. China has seen rapid development of high earth-rock dams thought it started construction relatively late. It is now at the forefront of the world in terms of the number and height of high earth-rock dams (200 m level) built and under construction. Examples include Tianshengqiao, Xiaolangdi, and Nuozhadu with designed 300-m-level high dams, Shuangjiangkou (314 m high), Rumei (315 m high), and Lianghekou (295 m high). Earth-rock dam construction technology will make a huge breakthrough.

Tianshengqiao Hydropower Project was officially launched in April 1991, realized water diversion on December 25, 1994, and put into operation the first generator in December 1998. The dam has a maximum height of 178 m, a crest length of 1137 m, and a crest width of 12 m. The reservoir submerged 4539 hectares of arable land and relocated 44,300 people. Xiaolangdi Hydropower Project is huge with construction spanning 11 years. It adopts an inclined core-wall rockfill dam with a designed maximum height of 154 m, a crest length of 1667 m, a crest width of 15 m, and a maximum width of 864 m. Upon completion, it inundates an area of 272.3 km^2 and controls a drainage area of 69.4 km^2. Earth amounting to 518.500 m^3 is used, and 1.2 m thick and 80 m deep, concrete cut wall is built, both setting new records in China.

The earth-rock dam for Nuozhadu Hydropower Station is 261.5 m high, the tallest of its kind in China and the third tallest in the world. It replaces the 160-m high Xiaolangdi Dam to be China's tallest dam by crossing the line of 100 m in height. As the theory, technology, experience, and specifications for dam construction applicable at that time cannot meet the construction requirements, the project systematically proposed, for the first time, a complete set of technologies for super high core-wall rockfill dams, which uses artificial gravel mixed with earth, as well as soft rock for rockfill materials. This encompasses the static and dynamic constitutive model for rockfill materials, the method for measuring hydraulic fracturing and fractures, a complete set of design criteria, and a comprehensive safety evaluation system for super high core-wall rockfill dams. Nuozhadu Hydropower Station has made and applied a number of innovative results with China's independent intellectual property rights, bringing China's rockfill dam construction technology to a new level.

4. Discussion

Hydropower development at the river basin level contributes to green and harmonious development by way of efficient use of hydropower resources.

Multi-objective cascade hydropower development was first proposed in Tennessee river basin development plan in 1933. The model was then successively implemented in the rivers of Cumberland, Missouri, Columbia, Colorado, and Arkansas after Tennessee in the United States. At the same time (1931–1934), the former Soviet Union drew up and put into practice the cascade development plan for the Volga River. In the next 40 years, fast progress was made in cascade development of hydropower. Most developed countries highlighted hydropower in energy strategies and exploited the majority of superior hydropower resources. While developed countries moved toward a steady period of hydropower construction in the 1970s, an upsurge with rapid cascade development took place in some Latin American developing countries in the 1960s. In the 28 years from 1958 to 1986, Brazil carried out a series of cascade development projects on the Paraná River and its tributaries, which encompasses 17 cascade power stations with a total storage capacity of 17.922 billion m^3 and a total installed capacity of 39.58 GW. This raised its world ranking in hydropower to 5th from 12th in 1950 with a scale of 1.54 GW.

The continuous deployment of hydropower projects not only provides a steady stream of power to ease the pressure on power supply but also drives economic development. However, dam construction along with water conservancy and hydropower projects has aroused some controversies. On the one hand, dams make abundant water available for agricultural irrigation that facilitates people's lives. Flood regulation and control by dams is also very important to largely avoid the loss of life and property. But on the other hand, dams slow down water flow, which easily leads to water pollution. The United States has begun to demolish some early built dams, and many problems have arisen during the demolition process, including the impact on the river basin and on the topography and geology. China has about 98,000 reservoirs and dams, the most among all countries. As this number increases, the impact on river ecosystems has drawn growing social attention. Such impact is manifested in two aspects. First, the ecological environment of rivers is fragmented by cascade development. For instance, there are 11 hydropower stations built and under construction in the 1326-km mainstream section of the middle and lower Jinsha River, and 10 large hydropower stations sitting on the mainstream alone of the 1060-km long Dadu River. Impoundment of these reservoirs and dams has changed the natural runoff and sediment transportation process of rivers, and especially water temperature. Fish migration channels have been blocked, affecting the survival and development of aquatic ecosystems to varying degrees. Second, the minimum ecological flow of rivers cannot be guaranteed. Construction of dams and reservoirs on rivers inevitably sparks conflicts between economic water use for water supply and power generation and ecological water use within rivers. If the relationship between household, production, and ecological water use is not handled well, the excessive emphasis on water storage to secure household and production water supply will often compromise the ecological flow of rivers. For example, there are many small and medium hydropower stations early built on small- and medium-sized rivers in southern China. They typically do not contain spillway facilities to discharge ecological flow and thus are unable to ensure sufficient ecological flow during the dry season. Diversion hydropower stations have even more impact on the ecological flow of rivers. Flow of small- and medium-sized rivers is naturally limited except during the flood season and varies widely between high- and low-flow periods. As a result of water diversion for power generation, flow is frequently deprived of the section between the barrage and the power station, bringing obvious damage to river ecosystems.

The concept of circular economy provides a new path for the development of hydropower. The circular economy has the characteristics of saving resources, protecting the environment, and promoting economic development, which coincides with the concept of sustainable development of hydropower. Circular economy mainly affects the power industry in two aspects: one is that circular economy can improve the conversion efficiency of energy and reduce the waste of natural resources; pollution of the surrounding environment caused by electromagnetic fields. As a kind of clean energy, hydropower has relatively little pollution to the natural environment but has a great impact on the ecological environment of the river basin. Hydropower stations should not only consider the benefits of power generation, but also comprehensive benefits such as shipping, flood control, and irrigation. How to improve the current operation mode of hydropower station on the basis of circular economy, so as to achieve the state of nature-society virtuous circle, there is still a lot of research space.

After more than 140 years of development, hydropower has received attention in many countries in the world and has become an irreplaceable and important part of today's clean energy. With the increase of the dam's operating time, its hidden problems have gradually emerged. Due to the different geographical environments and policies of different countries, countries have different ways to deal with the problems arising from the construction of new dams and the operation of old dams. However, there are still some problems that have not yet found a good solution, which has become a common problem faced by hydropower construction in the world. China's hydropower construction is at the world's leading level, and its dam construction technology, management, and operation methods are of great reference value for dam construction and for solving problems in dam operation.

5. Conclusion

In fact, humans have harnessed water for thousands of years. Due to the late invention of electrical technology, it was not until 1878 that the world's first hydropower station was constructed in France. In the next 100 years, hydropower gradually became the second largest source for power generation after thermal power by virtue of low operating cost, simple electromechanical equipment, and operational flexibility. Super hydropower stations born after 1990 are all from China, in contrast to foreign ones built before 1990. In particular, 11 of the world's top 20 super hydropower stations and 4 of world's top 5 hydropower stations are located in China. The Grand Coulee hydropower station from the United States, the oldest in the ranking, was commissioned in 1942 with an initial installed capacity of 1.97 GW, the largest at that time. It was expanded in 1967 and completed in 1980. China's Jinsha River, originating from the Qinghai-Tibet Plateau, renders the strongest power generation capacity secured by four super hydropower stations. This is attributed to large height difference, water abundance, and perfect terrain with towering mountains on both sides.

Along with rapid economic and social development, household and production water use has squeezed the natural runoff of rivers. Therefore, ecological water requirements that guarantee and maintain the stability of river and lake ecosystems are essential to the sustainable development of human society. Ecological flow is related to the life of rivers and lakes and considered an important indicator to express the ecological water requirements of rivers and lakes. Hydropower development imposes huge negative impact on the ecological flow of rivers. While the United States has begun to dismantle some of the dams built in early days, power generation will be the top priority in the future because the use of electricity as an alternative to oil will be multiplied amid the growing trend of substitution. In recent years, hydropower construction is recovering worldwide. Not only China has

DOI: http://dx.doi.org/10.5772/intechopen.103902

started construction of more than a dozen large-scale power stations at the same time, but also Africa and the Americas are building large-scale power stations.

However, hydropower station construction is an extremely expensive project involving a series of environmental, geological, and ecological issues. Past dam construction plans largely place emphasize on construction and operation phases and pay little attention to potential problems related to demolition and reconstruction. As the vision of ecological civilization has been widely accepted, it is increasingly recognized in recent years that rivers are a vibrant community of life with biological resource attributes such as water quantity, water quality, shoreline, hydropower, and aquatic organisms. Hydropower developers must not only integrate the conservation of river ecosystems as an important task in the planning, design, and construction phases but also assume responsibility for improving the quality and stability of river ecosystems during the operation phase. In this sense, a major research topic of hydropower development is to minimize the adverse impact of dams on the environment through environment-friendly reservoir construction and operation, and meanwhile, to make full use of reservoirs to rebuild the environment toward a sound situation of ecological improvement featuring harmony between man and water.

Acknowledgements

This work was supported by Beijing Natural Science Foundation of China (Grant No. JQ21029), and besides, the authors would like to thank Mr. Zeyang Zhao from North China University of Water Resources and Electric Power for doing a lot on the collection and collation of the relevant information and data.

Conflict of interest

The authors declare no conflict of interest.

Author details

Yizi Shang[1]*, Xiaofei Li[2] and Ling Shang[3]

1 State Key Laboratory of Simulation and Regulation of Water Cycles in River Basins, China Institute of Water Resources and Hydropower Research, Beijing, China

2 SANY Construction Technology Co., Ltd, Beijing, China

3 Nanjing Vocational College of Information Technology, Nanjing, China

*Address all correspondence to: yzshang@foxmail.com

References

[1] Kougias I, Aggidis GA, Avellan F, Deniz S, Theodossiou N. Analysis of emerging technologies in the hydropower sector. Renewable and Sustainable Energy Reviews. 2019;**113**:109257. DOI: 10.1016/j.rser.2019.109257

[2] Rauschenbach T. Evaluation and improvement of small hydropower plants in China. Wasser Wirtschaft. 2019;**109**(6):12-15. DOI: 10.1007/s35147-019-0092-9

[3] Sun LJ, Niu DX, Wang KK, Xu XM. Sustainable development pathways of hydropower in China: Interdisciplinary qualitative analysis and scenario-based system dynamics quantitative modeling. Journal of Cleaner Production. 2020;**287**:125528. DOI: 10.1016/j.jclepro.2020.125528

[4] Tang SW, Chen JT, Sun PG, Li Y, Yu P, Chen E. Current and future hydropower development in Southeast Asia countries (Malaysia, Indonesia, Thailand and Myanmar). Energy Policy. 2019;**129**:239-249. DOI: 10.1016/j.enpol.2019.02.036

[5] Wagner B, Hauer C, Habersack H. Current hydropower developments in Europe. Current Opinion in Environmental Sustainability. 2019;**37**:41-49. DOI: 10.1016/j.cosust.2019.06.002

[6] Kishore TS, Patro ER, Harish VSKV, Haghighi AT. A Comprehensive Study on the Recent Progress and Trends in Development of Small Hydropower Projects. Energies. 2021;**14**(10):2882. DOI: 10.3390/en14102882

[7] Mikhailov VE, Ivanchenko IP, Prokopenko AN. Modern State of Hydropower and Construction of Hydro Turbines in Russia and Abroad. Teploenergetika. 2021;**2**:5-15. DOI: 10.1134/S0040363621020041

[8] Yang SQ, Guo LL, Liu ZQ. Summary of the development of digital construction of water conservancy and hydropower projects. Journal of Hydroelectric Engineering. 2018;**37**(8):75-84. DOI: 10.11660/slfdxb.20180808. (in Chinese)

[9] Huang Q, Liu D, Wei XT, Yang YY. Analysis of the reasons why the number of dams built in China is the largest in the world. Journal of Hydroelectric Engineering. 2021;**40**(09):35-45. DOI: 10.11660/slfdxb.20210904. (in Chinese)

[10] https://www.globalgoals.org/zh

[11] Zhou X, Fan SY, Sun H, Tang L, Ma FW. Practices of environmental protection, technological innovation, economic promotion and social equity in hydropower development: a case study of cascade hydropower exploitation in China's Dadu River basin. Clean Technologies and Environmental Policy. 2021;**23**(10):2827-2814. DOI: 10.1007/s10098-021-02212-x

[12] Feng ZK, Niu WJ, Chen CT. China's large-scale hydropower system: operation characteristics, modeling challenge and dimensionality reduction possibilities. Renewable Energy. 2019;**136**:805-818. DOI: 10.1016/j.renene.2019.01.059

[13] Chu PH, Liu PK, Pan H. Prospects of hydropower industry in the Yangtze River Basin: China's green energy choice. Renewable Energy. 2018;**131**:1168-1185. DOI: 10.1016/j.renene.2018.08.072

[14] Sun YB. The achievement, significance and future prospect of China's renewable energy initiative. International Journal of Energy Research. 2020;**44**(15):12209-12244. DOI: 10.1002/er.5243

[15] Zhang BT. The Development of China's Hydropower and the Energy

Revolution and the Power Transformation. Journal of Hydroelectric Engineering. 2020;**39**(08):69-78. DOI: 10.11660/slfdxb.20200807. (in Chinese)

[16] Zhang CL, Chen SF, Qiao HJ, Dong LH, Huang Z, Ou CQ. Small hydropower sustainability evaluation for the countries along the Belt and Road. Environmental Development. 2020;**34**:100528. DOI: 10.1016/j.envdev.2020.100528

[17] Ansari MI, Saqib M, Agarwal P. Geometric Configuration Effects on Nonlinear Seismic Behavior of Concrete Gravity Dam. Journal of Earthquake and Tsunami. 2018;**12**(2):18500033. DOI: 10.1142/S1793431118500033

[18] Sevim B. Geometrical dimensions effects on the seismic response of concrete gravity dams. Advances in Concrete Construction. 2018;**6**(3):269-283. DOI: 10.12989/acc.2018.6.3.269

[19] Ganji HT, Alembagheri M, Khaneghahi MH. Evaluation of seismic reliability of gravity dam-reservoirinhomogeneous foundation coupled system. Frontiers of Structural and Civil Engineering. 2019;**13**(3):701-715. DOI: 10.1007/s11709-018-0507-1

[20] Sevieri G, De Falco A. Dynamic structural health monitoring for concrete gravity dams based on the Bayesian inference. Journal of Civil Structural Health Monitoring. 2020;**10**(2):235-250. DOI: 10.1007/s13349-020-00380-w

[21] Zhang XF, Yan TM, Liu Q, Zhang X, Wang L. Cracking analysis of induced joints in roller compacted concrete arch dam. Alexandria Engineering Journal. 2022;**61**(5):3599-3612. DOI: 10.1016/j.aej.2021.08.082

[22] Ahmadi MT, Soltani N. Mixing Regression-Global Sensitivity analysis of concrete arch dam system safety considering foundation and abutment uncertainties. Computers and Geotechnics. 2021;**139**:104368. DOI: 10.1016/j.compgeo.2021.104368

[23] Ding ZL, Zhu XY, Zhang HY, Ban HL, Chen Y. Analysis of the influence of asymmetric geological conditions on stability of high arch dam. Science and Engineering of Composite Materials. 2021;**28**(1):426-436. DOI: 10.1515/secm-2021-0043

[24] Cheng L, Wu YJ, Wei ZY. Summary and prospect of construction management and construction technology of super high arch dam. Water Power. 2020;**46**(10):73-181 (in Chinese)

[25] Duan B. Research on key technology of 300 m-level core rockfill dam. Northwest Hydropower. 2018;**1**:7-13 (in Chinese)

[26] Wang YK, Han MS, Li B, Wan YK. Stability evaluation of earth-rock dam reinforcement with new permeable polymer based on reliability method. Construction and Building Materials. 2022;**320**:126294. DOI: 10.1016/j.conbuildmat.2021.126294

[27] Chen SS. Key issues and countermeasures for the construction and safety of ultra-high earth-rock dams. Yangtze River. 2018;**49**(05):74-78. DOI: 10.16232/j.cnki.1001-4179.2018.05.015. (in Chinese)

[28] Yuan M, Qiang S, Xu YL, Li Y, Xu WQ. Research on Cracking Mechanism of Early-Age Restrained Concrete under High-Temperature and Low-Humidity Environment. Materials. 2021;**14**:15. DOI: 10.3390/ma14154084

[29] Tidke AR, Adhikary S. Seismic fragility analysis of the Koyna gravity dam with layered rock foundation considering tensile crack failure. Engineering Failure Analysis.

2021;**125**:105361. DOI: 10.1016/j.engfailanal.2021.105361

[30] Yang SL, Han XJ, Kuang CF, Fang WH, Zhang JF, Yu TT. Comparative Study on Deformation Prediction Models of Wuqiangxi Concrete Gravity Dam Based on Monitoring Data. CMES-Computer Modeling in Engineering & Sciences. 2022;**131**(1):49-72. DOI: 10.32604/cmes.2022.018325

[31] Liu PF, Chen JY, Fan SL, Xu Q. Uncertainty quantification of the effect of concrete heterogeneity on nonlinear seismic response of gravity dams including record-to-record variability. Structure. 2021;**34**:1785-1797. DOI: 10.1016/j.istruc.2021.08.098

[32] Beiranvand B, Komasi M. Study of the Arching Ratio in Earth Dam by Comparing the Results of Monitoring with Numerical Analysis (Case Study: Marvak Dam). Iranian Journal of Science and Technology-Transactions of Civil Engineering. 2021;**45**(2):1183-1195. DOI: 10.1007/s40996-020-00519-1

[33] Lyu WL, Zhang L, Yang BQ, Chen Y. Analysis of stability of the Baihetan arch dam based on the comprehensive method. Bulletin of Engineering Geology and the Environment. 2021;**80**(2):1219-1232. DOI: 10.1007/s10064-020-02009-0

[34] Chen JY, Qin LJ, Xu Q, Li J. Seismic Damage Indexes of a High Arch Dam Based on the Monolith. KSCE Journal of Civil Engineering. 2020;**24**(7):2063-2077. DOI: 10.1007/s12205-020-1202-z

[35] Alemie NA, Wosenie MD, Belew AZ, Kibret EA, Ayele WT. Performance evaluation of asphalt concrete core earth-rock fill dam relative to clay core earth-rock fill dam in the case of Megech Dam. Ethiopia. Arabian Journal of Geosciences. 2021;**14**:24. DOI: 10.1007/s12517-021-09009-8

Chapter 2

Hydropower in Russia: Case Study on Hydrological Management of the Volga-Kama Cascade

Pavel N. Terskii, Galina S. Ermakova and Olga V. Gorelits

Abstract

The capacity of hydroelectric power plants (HPPs) in the Russian Federation (RF) exceeds 50 GW. It is about 20% of the total capacity of all power plants in the country. The Volga River basin is the biggest in Europe with the catchment area of 1 360 000 km^2. It covers the most populated and most industrialized part of the European Russia. The largest cascade of reservoirs in Russia and Europe is the Volga-Kama cascade (VKC) constructed in 1930–1980. It consists of 12 great water reservoirs and HPPs with total capacity about 12 GW. The main peculiarity for the VKC management is the combination of different requirements by various economy sectors: safety, energy, navigation, water needs for domestic and industrial services, agriculture and fishery, recreation and ecological rules. These sectors often make conflicting demands for the VKC operation. The VKC management principle is to balance and satisfy all of them taking into account the changing climate and economical effectiveness. Modern decisions for the VKC management are based on two principles. First is the constant optimization of the whole VKC management rules, taking into account both climate change and the Strategy of the country development. The second is the constant technical modernization of the VKC equipment to achieve the best economical effectiveness and safety for ecosystems and population.

Keywords: Volga-Kama cascade, reservoirs, hydroelectric power plants, water resource management, water budget and regime, efficient operating, climate change

1. Introduction

Hydroelectric power plants (HPP) produce 16–17% of all electricity capacity in Russia. Currently in Russia there are [1]:

- 14 HPPs with installed capacity of over 1000 MW (6 of them in Volga river basin),
- 102 HPPs with installed capacity of over 10 MW,
- dozens of small HPPs.

Company "RusHydro" is one of the largest power generating companies in Russia. It is the leader in the generation based on renewable sources. "RusHydro" develops power generation based on the energy of water flow, sunshine radiation, wind power and geothermal energy [2]. There are several cascaded reservoirs constructed on the Great Russian Rivers – Angara and Yenisei cascade, Zeya and Bureya cascade, Lower Don cascade, Moscow river system cascade etc. The largest cascade of reservoirs in Russia is located on the rivers Volga and Kama.

The Volga River basin is the largest in Europe with catchment area about 1 360 000 km^2 (including Kama River basin). The total Volga River basin area is 40% of European territory of the Russia. The basin covers most populated and most industrialized part of European territory of Russia. It is populated with 58 mln. There are 7 cities with more than 1 mln population.

The Volga River is the longest river in Europe, its length from the source to the Caspian Sea is 3530 km. There are hundreds of tributaries along the main Volga River, the largest are Oka River (right tributary) and Kama River (left tributary). The Lower Volga region, including the unique ecosystems of Volga-Akhtuba Floodplain and Volga Delta – is the only natural part of the Volga River that is not affected by the backwater of anthropogenic HPPs dams. Annual average Volga water runoff (1881–2020) at the terminal gauging station "Volgograd" is 253 km^3. Maximum year runoff – 389 km^3 (1926), minimum year runoff – 160 km^3 (1975).

The largest cascade of reservoirs in Russia is the Volga-Kama cascade (VKC) constructed in 1930–1970s. The main reservoirs of VKC were fully completed by the beginning of 1960s. The VKC is the largest energy and transportation water system in the Europe (**Figure 1**).

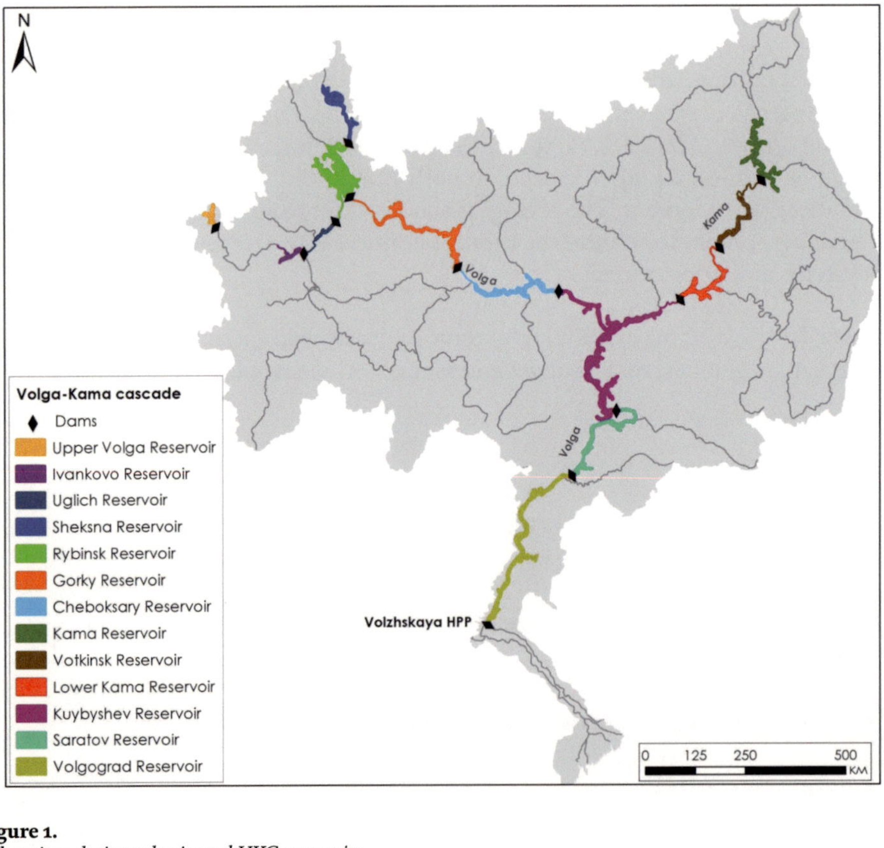

Figure 1.
Volga river drainage basin and VKC reservoirs.

Nowadays VKC includes 12 great water reservoirs with 12 Hydroelectric Power Plants (HPPs) and several small reservoirs without HPPs (**Figure 1**). Active storage of the VKC reservoirs is 80 km^3, total VKC storage – 175 km^3. Total design capacity of VKC HPPs is about 12 GW, actual capacity now is about 10,5 GW, annual hydropower generation – 35-40 billion KWh [2]. The HPPs of the VKC covering the peak part of the electricity consumption schedule are the backbone of the Unified Energy System of Russia, because it can increase electricity generation faster than other energy sources, such as nuclear or thermal power plants. The Volzhskaya HPP in the city of Volgograd – the downstream object of the VKC – is the largest HPP in Europe, with a total installed capacity of 2671 MW (**Figure 2**).

Construction of the Volga-Kama cascade was released as part of the great project "Big Volga", which was developed in the Soviet Union in the beginning of 1930s and was implemented from 1935 to 1960s. The "Big Volga" project assumed simultaneous solution of several serious problems of the European part of Russia economic development in 1930s: water transport, cheap energy, industrial and domestic water supply, agriculture and irrigation of arid regions, fisheries. There were several main purposes of the VKC construction. First was to create the transit waterway with guaranteed navigable depth about 4,5 m throughout the Volga River from upstream to the Caspian Sea, which connects the main industrial centers and raw materials regions. Second was to obtain huge amount of cheap energy. Third was the irrigation and industrial water supply.

Construction of three reservoirs in the upper stream of Volga river – Ivankovo, Uglich and Rybinsk – was the first step of the "Big Volga" project. It was started in 1930s - Ivankovo reservoir was built in 1937, but then the construction was suspended due to the Second World War. Uglich and Rybinsk reservoirs were completed only in 1955. The next step – creation of the Gorky, Kuybyshev, Kama and Volgograd (former Stalingrad) reservoirs – was fully completed by 1965 (**Figure 3**).

Figure 2.
Volzhskaya HPP in the city of Volgograd.

a)

b)

Figure 3.
Uglich HPP (a) and Kama HPP (b).

Votkinsk, Saratov, Lower Kama and Cheboksary reservoirs were built on the last stage of the construction.

Total head of Volga River from the source to the mouth – Caspian Sea – is about 256 meters. Total head of Volga River between the headwater of Ivankovo reservoir and tailwater of Volgograd reservoir is 135 meters, so it gives 8400 MW total actual capacity of the Volga hydroelectric power plants. Total head of Kama River between the headwater of Kama reservoir and tailwater of Lower Kama reservoir is 55 meters, so it gives 2150 MW total actual capacity of Kama hydroelectric power plants. The main characteristics of the VKC are shown in the **Table 1**, **Figure 4**.

Figure 5 demonstrates spatial heterogeneity of local catchment areas and local water inflow to reservoirs of the VKC. The 75% inflow is generated inside the biggest local catchments of Cheboksary, Kama, Lower Kama and Kuybyshev Reservoirs. Contribution of other reservoirs is less than 10% (for each of them).

№	Name of the reservoir (name of HPP if different)	River, lake	Year of completion	Reservoir full storage, *cub. km*	Reservoir active storage, *cub. km*	Time scale of regulation	Purposes*	Installed capacity of the HPP (actual capacity if different), *MW*
1	Ivankovo Reservoir	Volga	1937	1.22	0.89	daily	PNSFR	28.8 (25)
2	Uglich Reservoir	Volga	1955	1.25	0.67	interannual	PNSFIR	120
3	Sheksna Reservoir	Sheksna, Beloe lake	1964	6.51	1.85	seasonal	PNFR	84 (24)
4	Rybinsk Reservoir	Volga, Sheksna, Mologa	1955	25.4	16.7	seasonal	PNSFR	356
5	Gorky Reservoir	Volga	1961	8.82	3.90	seasonal	PNSFR	520
6	Cheboksary Reservoir	Volga	1981	4.60	0	seasonal	PNSR	1370 (820)
7	Kama Reservoir	Kama	1964	12.2	9.80	seasonal	PNSR	552
8	Votkinsk Reservoir	Kama	1966	9.36	4.45	seasonal	PNSR	1020
9	Lower Kama Reservoir	Kama	1979	4.21	0.77	weekly, daily	PNSR	1205 (566)
10	Kuybyshev Reservoir (Zhiguli HPP)	Volga, Kama	1959	57.3	30.9	seasonal	PNSFIR	2467
11	Saratov Reservoir	Volga	1971	12.9	1.75	weekly, daily	PNSFIR	1403
12	Volgograd Reservoir (Volzhskaya HPP)	Volga	1961	31.5	8.25	weekly, daily	PNSFIR	2671
13	Volga-Kama cascade	Volga, Kama	1845–1981	175	80.0	interannual, seasonal, weekly, daily	PNSFIR	11797 (10544)

**All of the VKC reservoirs are complex multipurpose impoundments, including following purposes:*
P - power generation;
N - navigation;
S - domestic and Industrial water supply;
F - fisheries;
I - irrigation;
R - recreation.

Table 1.
Main characteristics of the VKC [1, 3].

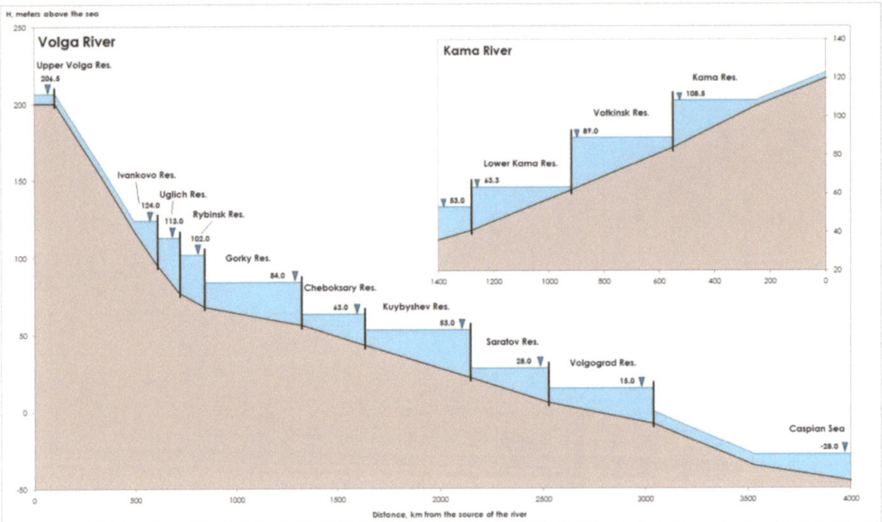

Figure 4.
VKC water surface profiles (based on materials [1, 4]).

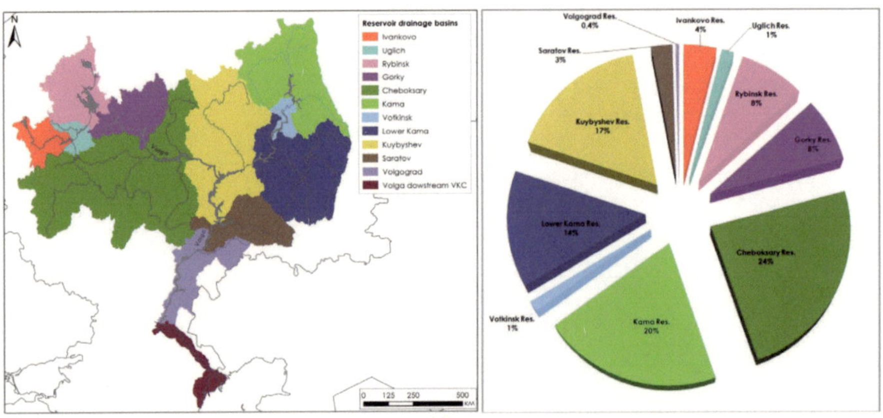

a) b)

Figure 5.
Local catchments of the VKC reservoirs (a) and contribution of the local inflow to the VKC total inflow, % (b).

2. Natural and manmade changes in the water budget and regime

Volga River basin and all reservoirs of the VKC are located on the southern slope of the East European Plain. The basin has some unique geographic features. From Middle Ages till present days Volga River system with its hundreds tributaries were located within the borders of one state – Russia. This feature distinguishes the Volga River basin from the basins of major European rivers and determines the features of its economical and cultural development. For example, Danube River basin partially covers the territories of 19 European countries, Rhine River basin – 6 countries, basins of the Dnieper, Daugava, Neman, Maas, Oder Rivers – partially covers the territories of 3 countries.

Volga River runs into the largest inland water body in the World – into the Caspian Sea. This unique feature defines a unique ecosystem of the river basin and the Sea and allows us to investigate the hydrological regime of this huge ecosystem in the framework of single hydrological cycle, which is determined by climatic changes, the anthropogenic influence and the VKC operation.

Climate changes in the Volga River basin – is the most important challenge of the last quarter of XX Century and the beginning of XXI Century. Against the background of a general increase in air temperature these increase in the European territory of Russia reached 0.5°C over 10 years. The main increase of temperature occurred in the winter period of the year, together with increase of humidity years [5, 6].

These climate changes significantly affect the water inflow to VKC. The temperature and humidity changes are resulted in the total annual inflow increase to VKC by 13% in last 40 years. Mean annual inflow to VKC for the period 1946–1977 is 248 km^3, while for the period 1978–2020 it is about 280 km^3 (**Figure 6**) [7, 8]. But between the sub-periods 1978–1995 and 1996–2020 there are significant positive changes in Evaporative Index end Dryness Index in the Volga basin, which caused a decrease in the inflow to the cascade from 295 km^3 (1978–1995) to 268 km^3 (1996–2020). The nowadays sub-period of 1996–2020 shows decreases in local inflow to VKC reservoirs and consequently decreases in water runoff through the Volzhskaya HPP into the unique ecosystems of the Lower Volga wetlands in the past couple of decades. Such climate-induced changes in the parameters of hydrological regime could explain only a part of the observed runoff decreases and changes in evaporative loss. The remaining, unexplained part is most likely related to the considerable changes in land-use at the agricultural regions of the Volga River catchment and other anthropogenic pressures [9].

Although climate changes much more strongly affects the seasonal inflow redistribution during last 40 years. The main feature of seasonal redistribution is alignment of intra annual unevenness of water inflow: great increase (about 63%) of winter water inflow because of warm and humid winters, increase (33%) of summer-autumn water inflow and slight decrease (2%) of spring flood period water inflow (**Figure 7**).

Significant intra annual water inflow redistribution caused the changes in the VKC operation. The winter water runoff and winter generation increased strongly.

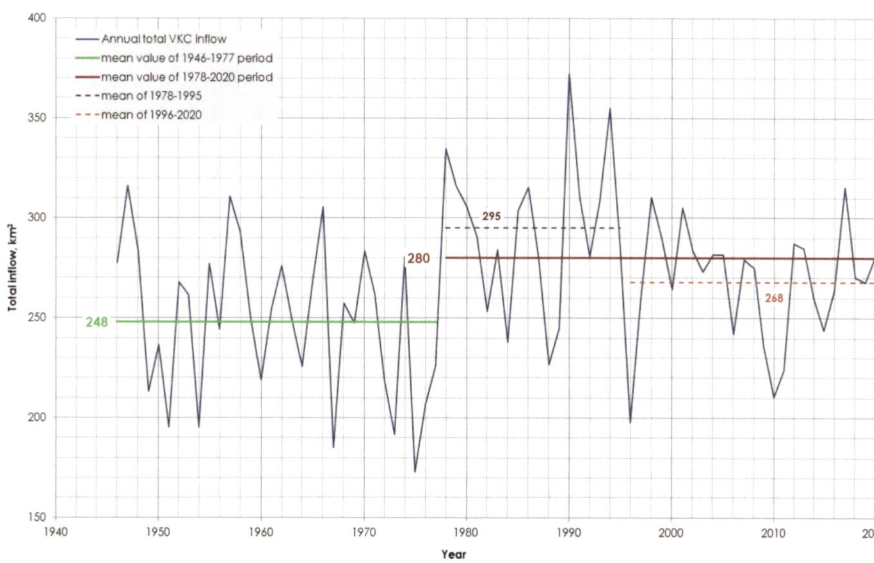

Figure 6.
Total annual inflow to the VKC (based on materials [7, 8]).

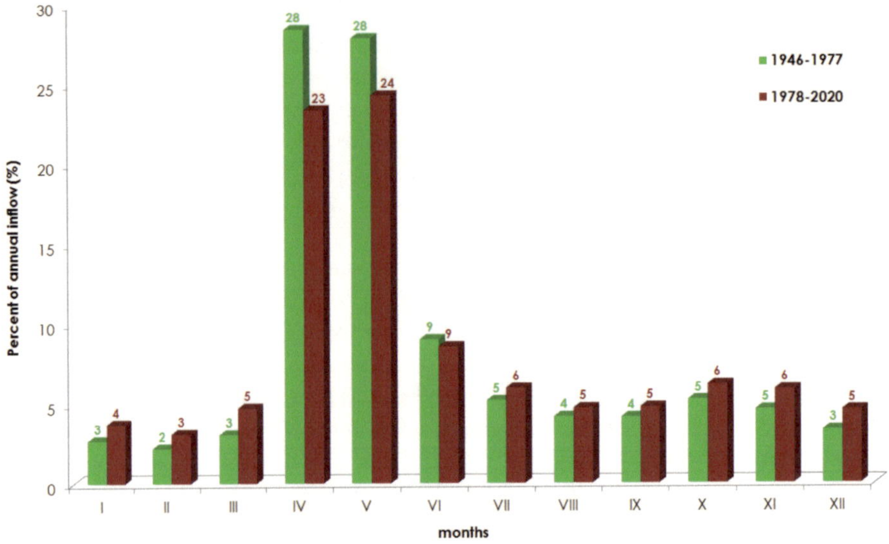

Figure 7.
Redistribution of the intra annual water inflow (based on materials [7, 8]).

3. Management principles

All of the VKC reservoirs are integrated-purpose water bodies. The complexity of the reservoir operation problem depends on purposes compatibility. If the purposes are more compatible - less effort is needed for coordination [10]. The main peculiarity for the VKC management is the combination of many requirements of various sectors of the economy: technical safety, energy, navigation, water needs for domestic and industrial services, agriculture and irrigation, fishery, recreation and ecological requirements. These sectors usually make conflicting demands to the VKC operation. The VKC management principle is to balance water users and consumers' demands and satisfy all of them taking into account the current conditions of a changing climate and the changed regime of local inflow to the VKC reservoirs and resulting total runoff downstream the cascade to the Lower Volga and Caspian Sea [5, 11].

There are several types of water consumption: domestic water supply, industrial and agricultural water consumption, irrigation. The largest water consumers are concentrated in the certain places associated with large cities. However, smaller ones are widely distributed along the lateral tributaries.

The important water users are hydropower engineering, water transport, fishery and recreation. To meet the requirements of these users, it is necessary to fill the reservoirs and not to exceed regulated water levels during the spring flood period. Reservoirs of the VKC serve to moderate both the flood risk and risk of summer droughts.

The reservoir operation is the water resource management, which regulates the water regime in interannual, seasonal, weekly and daily scales (sub-daily regulation also exists). The methods of reservoir operation are divided into operating curve method and rational operation [12]. Operating curve method makes possible the reservoirs functioning without detailed hydrological information and without enough range of hydrological forecast. The regulation procedures are being made depending on the operating curve, which represents the linkage between the day-by-day upstream water level and the discharge through the dam. The operating curve preparation is the goal of the water regime calculations based on stochastic programming for the optimal discharge planning.

Nevertheless, the operating curve cannot provide the reservoir regulation in case of emergency hydrological situations and in case of rapid change in the water management plans. Rational reservoir regulation begins to be implemented by applying the multiple forecast calculations of local inflow. In case of moderate situation (without rapid change in inflow forecast), rational regulation does not require excluding the operating curve from regulation. Rational regulation is prior to the operating curve method in case of complicated hydrological situation and changing inflow forecast.

This combination of the VKC regulation methods is based on the assessment of the forecast and hydrological situation (forecast-situational regulation). This kind of regulation is applied for different water seasons – spring and rain floods and flashes, summer and winter low water periods.

It begins with the hydrological forecast of the VKC distributed lateral inflow, compared to the water consumption plans and initial water level conditions for all reservoirs. Then different scenarios between the lowest and the highest water inflow are calculated and these results are used for the operation. If the real inflow begins to differ from the chosen plan, the scenario is being changed to fit the real inflow.

The main operating method for the low-water season is the compensatory regulation. Upstream reservoir storage gives the guaranteed discharge to provide the uninterrupted discharge through the downstream reservoir. The capacity of the VKC in the low-water season depends on the cascade regulation in the spring flood season. There are three main optimization tasks to operate the reservoirs: to choose the most rational regime of the spring flood transit, to choose the regime of water use during the low-water season or period, to optimize the cascade regulation using the operating curves. There is special software to meet this challenge. WATER RESOURCES and CASCADE systems serve the goal of choosing the VKC operating regime. ECOMAG is the distributed hydrological model with integrated operating curve method. It is used to calculate the distributed water inflow to the VKC [13, 14].

4. Problems and modern decisions

Multiple purposes of the VKC use leads to several types of conflicts between main users. According to [9] these conflicts are: conflicts in space, conflicts in time, conflicts in discharge. These conflicts have to be resolved in the most effective way by the reservoirs operation.

The main problems of the VKC operation are as follows:

- river runoff is the stochastic process with the significant calculation uncertainties;
- climate induced changes of annual and intra annual water inflow to the VKC, that are observed for last 40 years, require revision of current operating rules;
- the reservoirs capacity is limited with the construction and safety storage;
- multiple purpose regulation requirements meet conflicts among demands of various water users and consumers;
- during the low-water (and even average) years all water users and consumers are not ensured with water supply with the same high reliability.

To solve these problems the strategic and tactical planning has to be used. Strategic planning attended to solve the long term goal choice problem, i.e., the preferable realizable set of the values of water availability for its users and the key parameters of the VKC functioning. Tactical (annual) planning with decisions made at the level of Interdepartmental Working Group with the aim to take into account the specific hydrological and water-management conditions of the current year. These goals can be achieved by constructing the wide range of attainable probabilities. The final compromised decision is based on Pareto boundary for these multiple attainable probabilities by the consolidated negotiations between the main water users [14].

Modern decisions for the VKC management are based on two branches. First is the constant scientific-based optimization of the whole VKC management rules, taking into account both climate change and the Strategy of the country development.

One of the optimization perspectives is the change in the VKC operation by using the variable (against planned one) forecast-based water release in the spring flood period in the highest and lowest zones of inflow probability curve. It looks possible with help of long-range hydrological forecast (about 1 month). In case of estimated high water inflow reservoirs should release additional water volume. In case of low water forecast – reservoirs should decrease water discharge. It allows to decrease subsequent water volume, needed to fill reservoirs, and to increase water capacity for the Lower Volga supply. This kind of operation is mentioned for the pre-flood period in the Rules [14], but real methodology and legislative basis are not developed yet. Moreover it will require the revisions of the Rules.

The second is the technical modernization of the VKC to achieve the best effectiveness and safety. It should provide the required energy supply based on lower winter discharge. Filling the Cheboksary and Lower-Kama reservoirs up to projected levels can provide the ability to operate the effective and total storage of VKC in more efficient way [15].

5. Conclusions

The main environmental effect of river regulation in general and hydropower reservoir operations in particular is the alteration of the stream flow regime at various time scales, including seasonal, monthly, daily and sometimes hourly. This effect is fully manifested in the only natural area of VKC - the Lower Volga, including the unique ecosystem of Volga-Akhtuba floodplain, where the UNESCO Biosphere Reserve is located, and Volga Delta. The change in the spring flood regime led to a serious change in the landscapes and water bodies of the whole Lower Volga and the steppe formation of the northern part of Volga-Akhtuba floodplain. This negatively affected the state of the entire ecosystem of the Lower Volga, including Volga Delta and Northern part of the Caspian Sea [11].

The main results of VKC operation to the beginning of XXI Century are as follows:

1. Guaranteed navigable depths of about 4,5 m are provided throughout the Volga and Kama Rivers from upstream to the Caspian Sea.

2. Reducing the threat of floods and droughts are provided by VKC operation with seasonal, weekly, daily regulation.

3. VKC cheap electricity enters the Russia's Unified Energy System. VKC generation covers the peak loads in the Unified Energy System of Russia.

4. Water storage in VKC reservoirs provides sustainable water supply for the cities (population), industry (plants and factories), agriculture and irrigation, recreation and fishery.

5. Climate changes and water use protocols emphasize that reservoir operation should have flexible structure and scenario concept based on modern scientific and computational achievements.

VKC management and operation improvement is the governmental goal of whole institutes. Authors only may suggest some general vectors to solve main problems of the VKC operation. First - to increase the density of hydrological gauging network in the VKC catchment to have the basis for improvement of the stochastic description of the water inflow. Second - to develop more effective medium and long range hydrological forecasts. Because of problem of water inter annual and long-term temporal distribution these improved forecasting methods should support the decreasing of water losses during flood periods and save it for the deficit periods by dynamical storing it in the VKC. Finally - to develop the legislative basis for new implementation of variable operating rules of water management of the VKC.

Acknowledgements

The chapter is based on materials and data collected under the State contract #10-GK/FCP-2013 "Water management complex of the Russian Federation in 2012-2020" realized by Zubov State Oceanographic Institute (SOI) with significant contribution of Russian Academy of Science Water Problem Institute (WPI), Kostyakov All-Russian Research Institute for Hydraulic Engieneering and Land Reclamation (VNIIGIM), Caspian Research Institute for fishery, Hydrometeorological Research Center of Russian Federation (Hydrometcenter).

Conflict of interest

The authors declare no conflict of interest.

Notes/thanks/other declarations

Authors are thankful to members of Automated Methods lab in SOI and personally to Igor Zemlyanov (PhD, head for the lab, vice director of the Institute) for the overwhelming support, Aleksander Buber (VNIIGIM), Mikhail Bolgov (WPI) and Vladimir Kryzhov (Hydrometcenter) for consultations.

Author details

Pavel N. Terskii[1]*, Galina S. Ermakova[2] and Olga V. Gorelits[2]

1 Faculty of Geography, Lomonosov Moscow State University, Moscow, Russia

2 N.N. Zubov State Oceanographic Institute, Roshydromet, Moscow, Russia

*Address all correspondence to: pavel_tersky@mail.ru

References

[1] Dvoretskaya M I, Zhdanova A P, Lushnikov O G, Sliva I V. Handbook of Renewable energy. Hydroelectric power plants of Russia. St. Petersburg: Polytechnic Publishing House; 2018. 224 p.

[2] The official website of the “RusHydro” company. Available from: http://www.eng.rushydro.ru/company/in_brief/ [Accessed 2021-04-21]

[3] Georgievsky V Y, editor. Scientific and applied reference book: Long-term characteristics of water inflow into the largest reservoirs of Russian Federation. Moscow: RPC Ofort; 2017. 132 p.

[4] Kryzhov V N, Gorelits O V. The Arctic Oscillation and its impact on temperature and precipitation in Northern Eurasia in the 20th Century. Russian Meteorology and Hydrology. 2015; 40: 711-721. DOI: 10.3103/S1068373915110011

[5] Juzbaši A, Kryjov V N, Ahn J- B. On the anomalous development of the extremely intense positive Arctic Oscillation of the 2019-2020 winter. Environment Research Letters. 2021; 16, 05500. DOI:1088/1748-9326/abe434

[6] Georgievskii V Y, Alekseev L P, Litova T E, Dubrovskaya K A, Zadonskaya O V, Fuksova T V. Water inflow into the reservoirs of the largest hydroelectric power plants in the Russian Federation for different time intervals. Database Official Registration Certificate RU 2016621140, 22.08.2016 (In Russian)

[7] State report “Water resources conditions and use in Russian Federation in 2015”. Available from: https://water-rf.ru/water/gosdoc/378.html [Accessed: 2021-04-25]

[8] Ahlén I, Vigouroux G, Destouni G et al. Hydro-climatic changes of wetlandscapes across the world. Sci Rep. 2021; 11, 2754. DOI: 10.1038/s41598-021-81137-3

[9] Jain S K, and Roorkee NIH. Introduction to reservoir operation. Technical Report. 2019.

[10] Gorelits O V, Ermakova G S, Terskii P N. Hydrological regime of the lower Volga river under modern conditions. Russian Meteorology and Hydrology. 2018; 43, 10: 646-654. DOI:10.3103/S1068373918100035

[11] Levit-Gurevich L K. Rational water resources management of reservoirs on the example of Volga-Kama Cascade. Izvestia of Samara Scientific Center of the Russian Academy of Sciences. 2012; 14 (4): 2343-2354.

[12] Danilov-Danilyan V I, editor. Justification for water management strategies. 2006. 336 p.

[13] Motovilov Y G. ECOMAG: a distributed model of runoff formation and pollution transformation in river basins. IAHS Publ. 2013; 361: 227-234.

[14] Bolgov M V, Buber A L, Komarovskii A A. Searching for Compromise Solution in the Planning and Managing of Releases into the Lower Pool of the Volgograd Hydropower System. 1. Strategic Planning. Water Resources. 2018; 45: 819-826. DOI: 10.1134/S0097807818050044

[15] Draft of the Rules for the Use of the Kuybyshev, Saratov, Volgograd Reservoirs. Federal Agency of water resources: Moscow, Russia, 2012. Available from: http://nvbvu.ru/files/4517.pdf [Accessed 2021-08-26]

Section 2

Hydropower Technological Innovations

Chapter 3

Hydro Power Tower (HYPOT)

George Mamulashvili

Abstract

Humanity has used the power of falling water for centuries to produce electrical energy, but there have been no significant changes in technology. Marine Energy has received an explosive development. Traditional technologies are passive and have low efficiency. It is not possible to use the effect of falling water in the ocean. The chapter considers the technology, which allows to convert not only the kinetic energy of a moving horizontal flow, but also the potential energy of water hammer in a combination of pressure drop between layers of water that have different hydrodynamic characteristics. This is a high efficiency due to the use of the Pitot-Prandtl tube principle and Bernoulli's law and in combination with the effect of raising the water of the hydraulic ram. The calculations are based on computational fluid dynamics (CFD) methods. It is known that 94% of incoming solar energy is converted into underwater currents and only 6% - on the surface. Therefore, the proposed technology can be highly competitive in relation for example to Orbital Marine Power (OMP) project and another known offshore wind and wave power plants which convert only the kinetic energy of the surface air and sea currents.

Keywords: Marine Energy, Water hammer, Shock wave, Pulsating flow, Pitot tube

1. Introduction

The formulation of the project theme includes general information about the facility as a new source of renewable energy for the ocean.

This is an underwater gravitational energy technology, which is one of the most promising generating devices due to the significant potential of generating electrical energy, as it converts a large volumetric part (almost 94%) from all the potential solar energy captured by the oceans.

A hydroelectric power plant perceives the kinetic energy of currents and the potential energy accumulated by it due to water hammer and pressure drop between the layers. It artificially creates a rising whirlpool in the open sea. At the same time, the gift wave from the water hammer propagates through the two-phase hyperbolic project HYPOT and increases the pressure - in the positive direction, when falling - in the negative direction. This occurs when there is a sharp change in the direction of the current in the neck of the tower. The destructive effect of this phenomenon is associated with the inability of the fluid to contract smoothed out by the hyperbolic curve.

The chapter presents the main assumptions and results of the calculation of the digital twin, as well as the design methods of the HYPOT project. **Figure 1** shows a general view of the hydroelectric power plant of the cyclone action.

The HYPOT project in the complex can convert the kinetic energy of tidal and bottom flows, as well as the potential energy of pressure drop at different salinity

Figure 1.
Underwater hydroelectric power plant of cyclone action.

and water temperature. With the help of water hammer, the kinetic energy of the moving liquid is transferred into the potential energy of the resting liquid. However, such a transition is not instantaneous, but proceeds at a certain speed, depending on the properties of the liquid and the geometry of the pipeline. The HYPOT enclosure has a two-phase hyperbole geometry that reproduces the narrowing configuration in the center of the torus.

This is done by analyzing the vector of motion of the lifting flow for the maximum approximation to natural conditions. With the tower version, the tower creates the initial necessary pressure for the operation of pulse devices (on the principle of "water hammer"), so the project refers to a gravitational-pulse hydroelectric power plant, since the potential energy of water is the gravitational energy accumulated in it.

2. Theoretical prerequisites for calculation

Calculations of the HYPOT prototype in the ANSYS software package clearly proved the effect of water hammer into the neck of the tower on the increase in flow [1–3]. **Figure 2** shows spatial scheme of HYPOT digital twin for calculation in ANSYS as opposed to simple OMP [4]. The diagrams in **Figures 3** and **4** show how the pressure vector increases as the current in the collector moves to the neck of the tower, where there is a sharp pressure drop of 26.5 times, and the jump in the value of the flow vector increases respectively to 87.54 m/s due to water hammer. The calculation is made with the assumption that the entire volume of the incoming water flow flows into the collector. In order to find the balance of the incoming water into the collector and bypass it, it will be necessary to further solve the problem of multipoint calculation of the hydroelectric power plant, including the maximum possible sphere of water surrounding the station, in order to understand the losses at the entrance to the collector. Since the station works in general with water hammer and the release of water through the upper nozzle, the assumption that the entire volume of water will fall into the collector has a small error due to strong centripetal and forward motion along the current upwards in a hyperbolic tower.

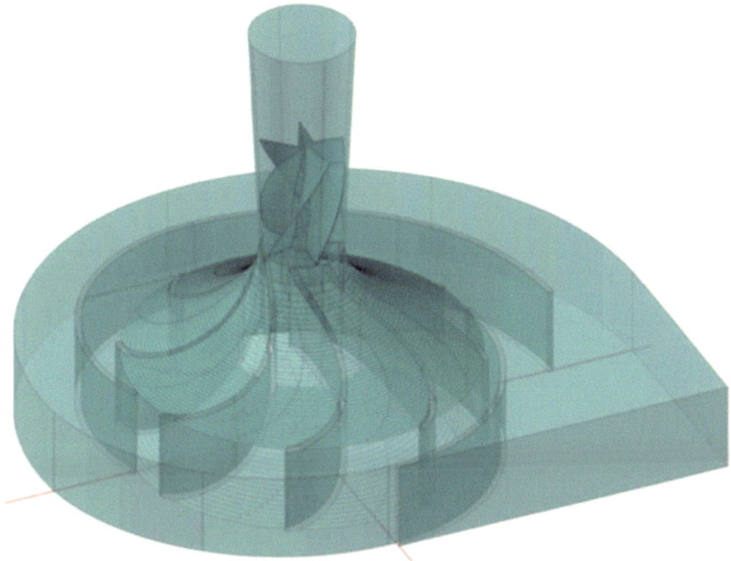

Figure 2.
Digital twin.

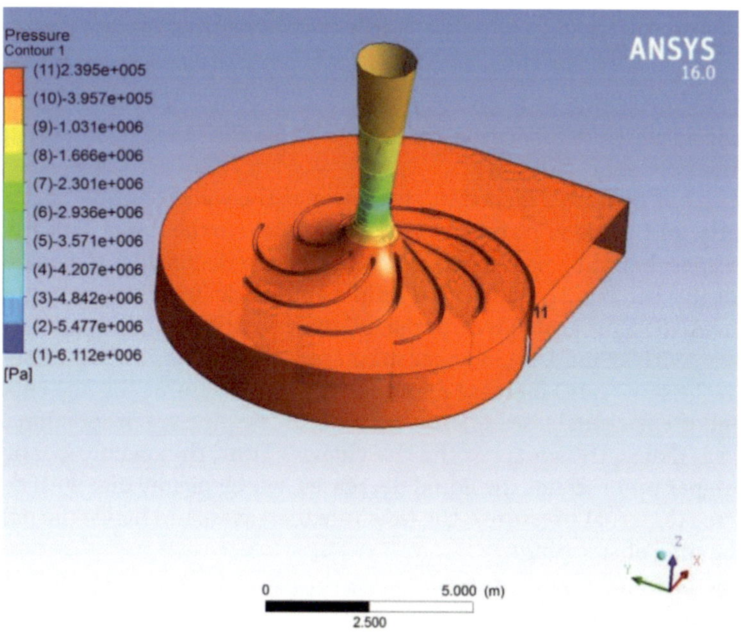

Figure 3.
Pressure at the speed of 4.5 m/s and inlet water flow of 32,4 m³/s.

Hydraulic shock at HYPOT is a short-term, but sharp and strong increase in pressure in the collector with a sharp braking of the fluid flow moving through it from the outside. The phenomenon of water hammer [5] here is creative - it is with its help that an impulse is given to the water intake, which then obeys Bernoulli's law of communicating vessels rises up, throwing water from the nozzle of the tower under high pressure.

First of all, it is necessary to take into account the high speed of the water hammer process. Since the speed of movement of the boundaries of zones with different

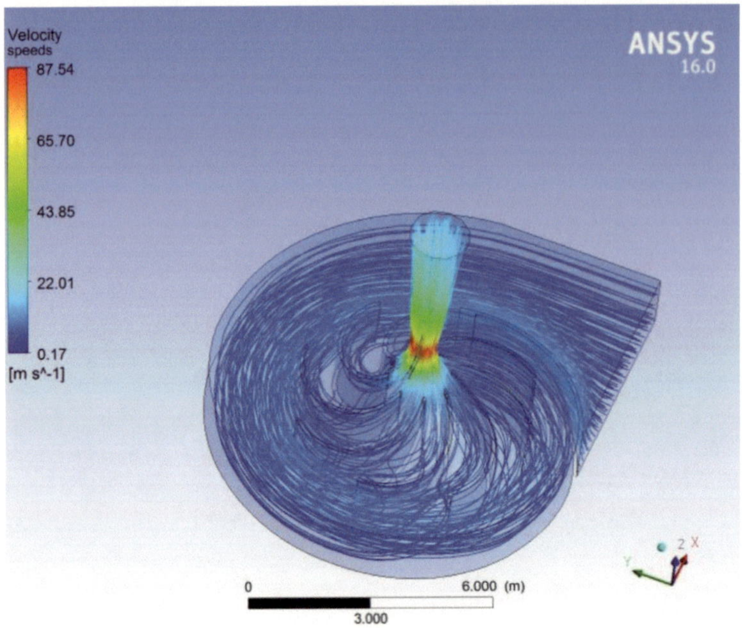

Figure 4.
Volocity speeds at the water speed 4.5 m/s and the incoming water flow of 32.4 m^3/s.

pressures at high rigidity of the body and neck is determined by the speed of propagation of elastic deformations in the liquid, i.e. the speed of sound, everything happens in a very short time.

As the size of the tower increases, the power of the water hammer increases significantly, and at the same pressure at the entrance to the tower, this growth is usually steeper than the linear dependence. Here we will consider the qualitative reasons for this behavior (quantitative results automatically follow from the calculations in the ANSYS program given in the following sections of this page).

However, with an increase in linear mass sizes (and, consequently, kinetic energies at the same rate) increase in proportion to the volume, i.e. the cube of their change, and the friction losses against the walls of the pipe are proportional to the contact area, that is, the square of the size change. Thus, the specific loss of energy per friction per unit mass of the liquid decreases, which means that with the same driving force (external pressure), the flow rate increases, and hence the pressure jump at the time of stopping.

It should be noted that the pressure jump during water hammer does not depend on the initial pressure that caused the liquid to move through the tower, but depends only on the speed obtained by it. This means that the acceleration of a liquid with a relatively high pressure in a short time can be replaced by a longer acceleration under the influence of lower pressure. However, it will not be possible to indefinitely reduce the acceleration pressure: first, in real conditions, the low pressure already at a not too high flow rate will all go to compensate for hydraulic friction; secondly, even for super fluidity, there is a limit to the maximum speed that the flow can reach at a given head at the entrance to the tower in accordance with Bernoulli's equation.

However, it is this circumstance that allows hydraulic rams to raise the fluid to a height many times higher than the difference in levels that leads them.

Finally, it should be noted that the vacuum, up to the almost complete absence of pressure with a strong water hammer, does not mean that at this stage the liquid leaves the entire tower pipe. This only means that the liquid ceases to put pressure on its walls. In reality, the void is formed only in the separation zone near the neck of the tower - in the same place where there was a water hammer with a sharp change in flow.

Where does the fluid accelerate?

First of all, it is necessary to find out where the acceleration of the liquid occurs - in the tower or outside it? The continuity equation gives an unambiguous answer: inside the tower of the unchanged cross-section, the flow rate is also unchanged, which means that all the acceleration occurs in the tank in front of the tower! It is easy to imagine by observing the discharge of water from the bath - the "funnel" over the drain hole is due to the zone of acceleration of water, which is located in the volume of the bath itself, and in the drain pipe the water speed no longer changes. Therefore, the water hammer energy is due to the fact that the entire volume of water moves in the pipe at the same speed.

Involving fluid in motion outside the tower.

Involving the fluid filling the tower in the movement beyond it.

The paler color in the chart shows areas at a higher rate. Gradations are shown conditionally, the increase in speed is sharp.

Shock wave damping [6].

As the liquid accelerates before entering the tower when the fluid in the collector has stopped as the result of water hammer, the liquid that has already gained some speed near the manifold entrance is forced to stop. This stop causes an increase in pressure around the inlet to the tower, which is often interpreted as "shock wave exit from the pipe".

However, the pressure drop is large, and therefore the liquid moves faster. Then the pressure outside the tower drops rapidly, and the speed of movement of the liquid outward also increases rapidly.

Finally, it should be recalled that all the processes described here occur very quickly in microseconds!

Above we have considered the water hammer from the "traditional" mechanistic positions.

It should be noted that for a short time, water hammer puts the substance in extremely extreme conditions - the pressure can increase by hundreds or even thousands of atmospheres, which corresponds to conditions at a depth of tens of kilometers. But even if the pressure does not grow very much (by dozens of atmospheres, or even just by several atmospheres), the rate of pressure changes for each particle of matter that falls under the influence is very high - 1012 Pa/s or more. It is quite comparable, and even exceeds the rate of change in pressure during explosions. At the same time, the gas or plasma environment formed during explosions is very compressible - it "absorbs" the impact, and a little further from the epicenter the pressure rises much more smoothly. But during water hammer, due to the low compressibility of liquids and the high rigidity of the wall material, this ultra-fast pressure jump affects almost the entire volume involved in the water hammer. Such sharp jumps in pressure correspond to gigantic accelerations and inhibition of particles of matter when the shock wave front passes through them. True, they last nano- and picoseconds, so the total displacement of liquid particles is small and usually is, in accordance with its low compressibility, micrometers or nanometers. However, by the standards of atoms and molecules, these shifts are very large, and the resulting forces are also significant.

For example, Carré (1705) observed a curious phenomenon: a bullet fired into a wooden box filled with water exploded. A shock bullet, transmitting a large pulse to the water, generates a shock wave that tears the walls [6].

3. Analytical calculation

The subject of these applied research and experimental developments planned for the project is, first of all, the determination of the forces of intermolecular interaction of water in the stream at different pressures and ambient temperatures and when using a cyclone amplifier. Ocean currents carry kinetic energy obtained from solar radiation, entering the collector, the current experiences a sharp drop in pressure on the rise into the neck of the tower and increases the speed due to water hammer, which closes the chain reaction of overcoming gravity and ejecting water through the nozzle of the tower.

Depending on this, the flow rate and volume are calculated to generate electrical energy in a two-phase hyperbolic housing by a spiral turbine, which ultimately

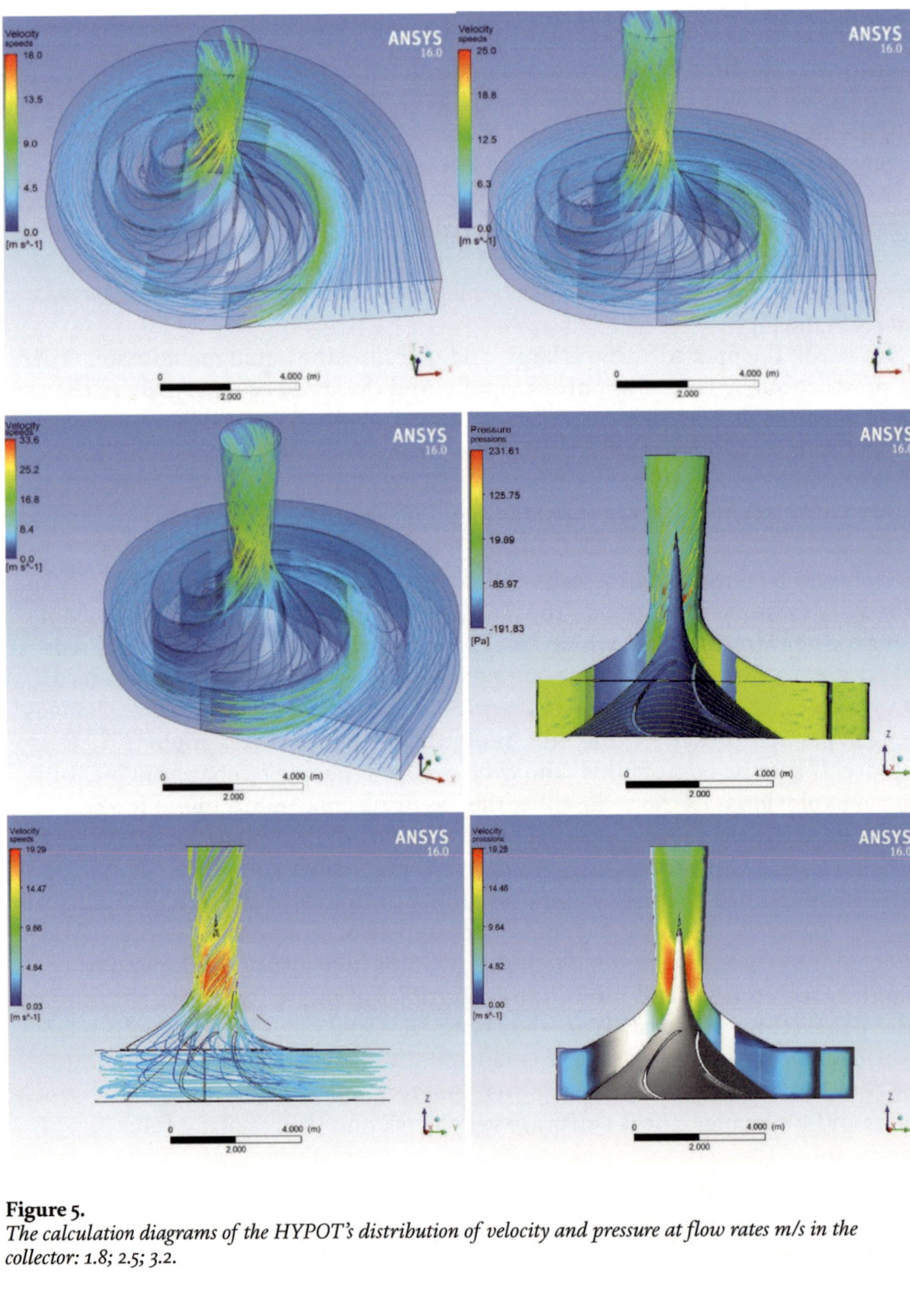

Figure 5.
The calculation diagrams of the HYPOT's distribution of velocity and pressure at flow rates m/s in the collector: 1.8; 2.5; 3.2.

z	7,5					
R (M)	Vz	Wabs	Vt	Beta	phi	h jet (m)
0,2	-0,87	1,28	0,94	−42,82	−0,93	0,014
0,3	1,57	2,66	2,15	36,17	0,73	0,044
0,45	4,87	6,22	3,87	51,53	1,26	0,423
0,5	5,75	7,19	4,32	53,10	1,33	0,590
0,7	9,17	10,91	5,91	57,19	1,55	1,501
0,9	10,87	12,85	6,85	57,77	1,59	2,109
z	3,5					0,219
R (M)	Vz	Wabs	Vt	Beta	phi	h jet (m)
0,2	13,37	19,15	13,71	44,28	0,98	3,190
0,3	13,58	18,13	12,01	48,51	1,13	3,291
0,45	12,8	17,2	11,49	48,09	1,11	2,924
0,5	12,7	16,56	10,63	50,08	1,20	2,878
0,7	11,85	14,83	8,92	53,04	1,33	2,506
0,9	10,87	12,85	6,85	57,77	1,59	2,109
Z	4,6					0,378
R (M)	Vz	Wabs	Vt	Beta	phi	h jet (m)
0,2	8	13,27	10,59	37,08	0,76	1,142
0,3	10,35	14,78	10,55	44,45	0,98	1,912
0,45	11	14,8	9,90	48,01	1,11	2,159
0,5	10,8	14,53	9,72	48,01	1,11	2,081
0,7	9,36	12,82	8,76	46,90	1,07	1,563
0,75	9,83	12,5	7,72	51,85	1,27	1,724

Table 1.
Results of preliminary calculations of the tower for the HYPOT project.

determines all energy production. Based on the effect of Italian physicist Giacomo Batista Venturi, Daniel Bernoulli Low, Henry Pitot tubes [7] and the Navier–Stokes equation for incompressible liquid, using ANSYS software for the hydropower tower calculation scheme. **Figure 5** shows the calculation diagrams of the HYPOT's distribution of velocity and pressure at flow rates m/s in the collector: 1.8; 2.5; 3.2.

Preliminary calculations of the tower at a depth of 30 meters showed the following results, which are summarized in **Table 1**.

4. Testing the digital twin of the lower

Based on the preliminary calculations given in Chapter 2, the international HYPOT project developed a prototype of a digital twin hydroelectric power plant for installation in the Strait of Messina off the coast of Sicily (Italy).

The international project included the results of the calculation of an underwater hydroelectric power plant with a tower height of 7.5 m. Below is **Figure 6** of the HYPOT's section of the power plant developed as part of the project.

As you can see from the diagrams above, the initial flow is not essential. for generated hydraulic energy. The main role is played by the pressure difference between the layers and the configuration of the intake manifold, which provides conditions for the occurrence of water hammer and obtaining the strongest acceleration in the neck of the tower. In addition, various sections from round to elliptical were tested from view of the analysis of the hydraulic power of the plant and the results are summarized for the selection of tower sections **Figure 7**. These graphs, being a purely empirical document, should not be distributed in one form or another, in addition, they are valid not only for choosing the configuration of the tower section.

The diagram below in **Figure 8** shows the kinetic energy levels available in the tower. We can see that 41% of this energy is still present at the exit of this tower, the rest is spent on walking from the pass to the exit. This means that up to 41% of the total energy entering the tower can be used to convert into a vortex turbine (the results are deposited from ANSYS CFD). The red curve is something that would be desirable to implement with a turbine so that it can return the maximum energy obtained in both images.

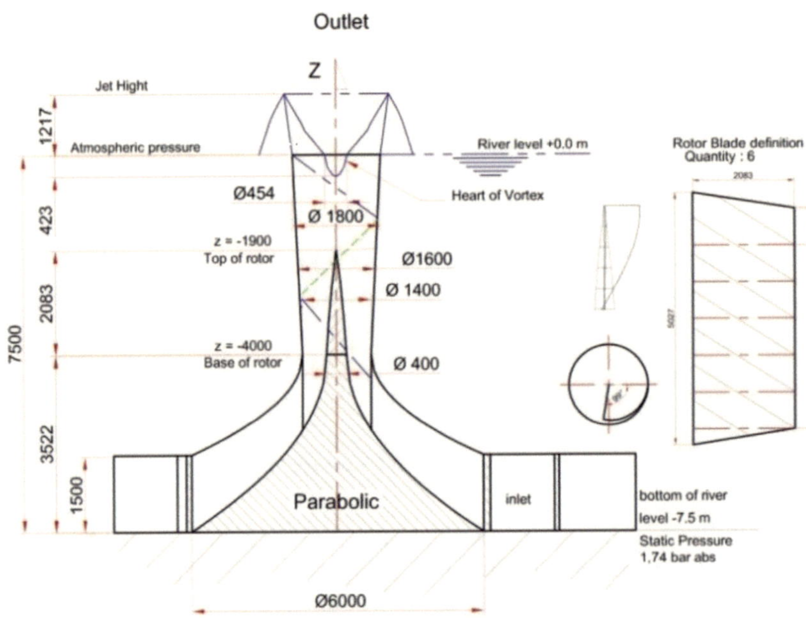

Figure 6.
The HYPOT's section of the power plant.

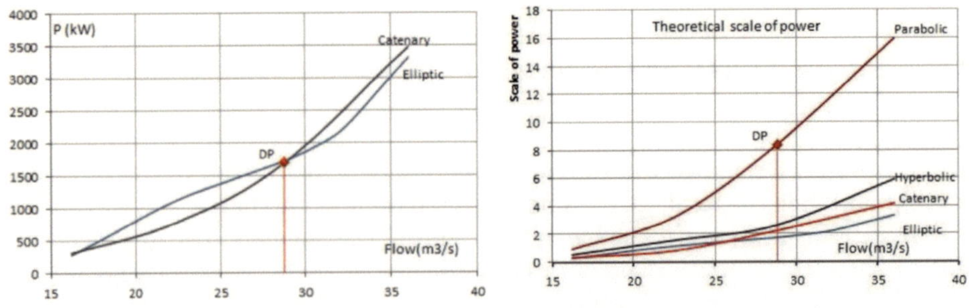

Figure 7.
Analysis of hydraulic power.

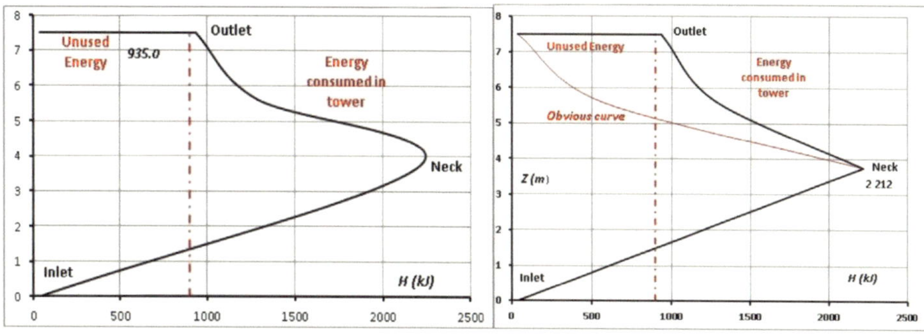

Figure 8.
Analysis of kinetic energy levels.

Figure 9.
Prototype of the HYPOT model in 1:3 scale.

The prototype of the HYPOT's model of scale 1:3 is designed to test the principle of operation of the entire system in the conditions of the mouth of the river flowing into the open sea, shown in **Figure 9**.

The subject of these studies is the problem of creating a new technology for the stream generation of powerful products for underwater hydroelectric power plants and hydrogen production services. The subject of the project is current scientific research (theoretical and experimental), as well as the development of an experimental technical and technological solution for the production of electrical energy in an artificial whirlpool with the possibility of obtaining hydrogen to replenish the peak load of the power plant and use oxygen waste to clean the polluted ocean.

Thus, the subject of the application reflects the research essence and nature of the work (subject and object).

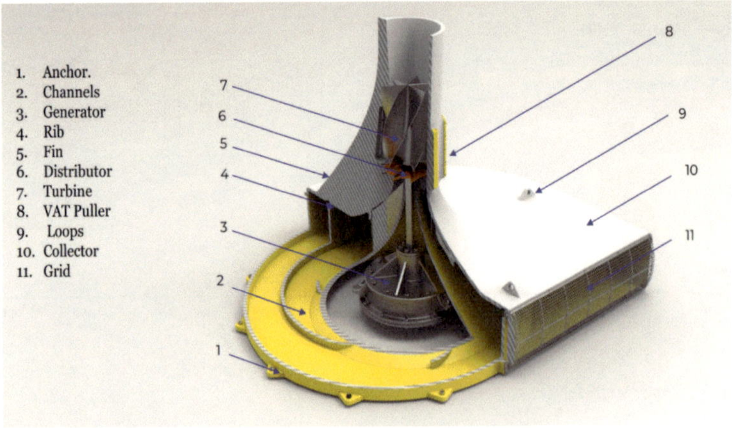

Figure 10.
Industrial project of HYPOT.

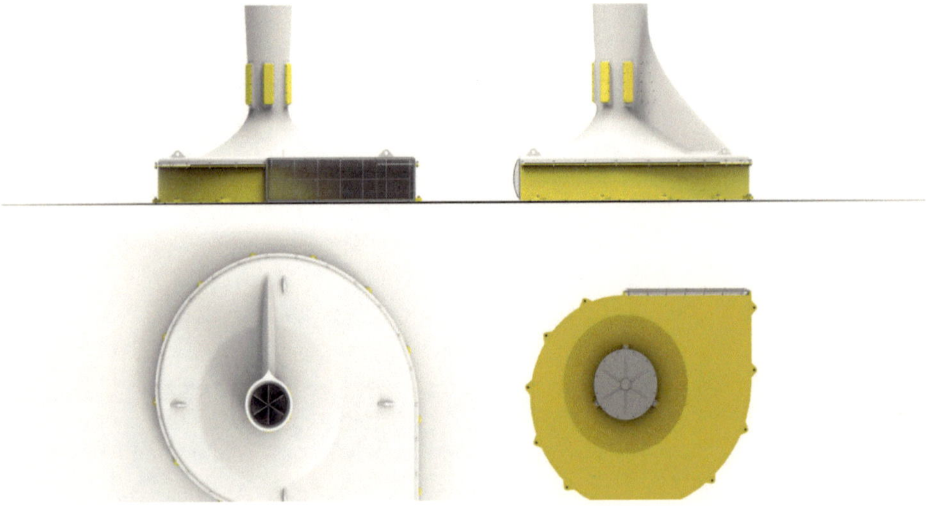

Figure 11.
Industrial prototype of the HYPOT in the Cartesian coordinate system.

In the following sections, the wording to the description of the subject of the proposed work, as well as the characteristics of the composition of the work and the scientific and technical results of the work on the proposed project, contain the planned innovative solution of various bases (sea suspensions on the pontoon and river installed on the bottom at the mouth of the rivers when they fall into the ocean), which determines the image and contributes to the creation of the future product, which in turn is the determining condition for the implementation of the Horizon 2020 Framework project. Similar innovative marine renewable energy technologies and their integration into the energy system of the European Union, call to the Building low-carbon, climate-resilient future based on unique High-performance technologies [8].

The industrial prototype of the hydropower tower an as underwater hydro-electric power plant, including a collector with a protective grid, a generator on permanent magnets, a vortex turbine, a tower, a steering bar with the possibility of turning downstream shown in **Figure 10**.

Initial assumptions for the calculation of the prototype: The consumption in the design of the HPT prototype is taken 2 m/s, and the water consumption is 18,000 kg/s. The diameter of the neck of the tower at Z = 3.5 m - 1400 mm, Z = 5.6-1600 mm, Z = 7.5-1800 mm.

Preliminary laboratory research work of a hydropower tower model showed that there is a correlation between the power emitted by the jet nozzle and the distance to the surface of the water. That is, the lower the underwater power plant is installed, the higher its power should be for the stability of the entire complex.

The main elements of the model of the underwater hydroelectric power plant of the HPP were made of composite materials and painted with water-resistant nitro paint, since the main condition was to test the high corrosion resistance of the station to ensure its long-term use under water.

The HYPOT in the Cartesian coordinate system shown in **Figure 11** is designed to test the principle of operation of the entire system and compare theoretical and experimental results.

5. The subject of the research

The subject of the research is the problem of creating a new technology for the stream generation of high-power products for underwater hydroelectric power plants and hydrogen production services. The subject of the project is topical scientific research (theoretical and experimental), as well as the development of an experimental technical and technological solution for the production of electrical energy in an artificial whirlpool with the possibility of obtaining hydrogen to replenish the peak load of the power plant and use oxygen waste to clean the polluted ocean.

It is planned to apply a fairly simple method of dissociation of water into hydrogen and oxygen and a device for its implementation, suitable for industrial use, which will reduce the energy intensity of the water dissociation process and ensure the possibility of separate production of gases.

To solve the problem and achieve the claimed technical result with a known method of dissociation of water for hydrogen and oxygen, including the effect of an electric field on water or water electrolyte through electrodes located at a distance from each other, and the removal of dissociation products, the effect on water or electrolyte of water by an electric field is carried out with a calculated resonant frequency on harmonics, in relation to which the frequency of natural oscillations of water molecules is multiple. And the dissociation products are removed separately from each even and odd electrode.

Of course, the project will use publicly available data from the experience of construction and operation of all known underwater hydropower projects.

The proposed design of an underwater hydroelectric power plant with a vertical turbine and a hyperbolic housing is very different from conventional wind turbines immersed in water.

Unlike the Orbital Marine Power [4], the "Sea Gen" [9] and another invention [10], HYPOT has a steering stabilizer that easily deploys the structure in the direction of the current, which does not require additional expensive equipment to track the direction of the tidal current, which significantly reduces the construction of an underwater hydroelectric power plant.

The steel structure of the hydroelectric power plant is firmly fixed on the seabed on stilts.

It is necessary to compare the cost of building the most powerful offshore wind turbine and a small HYPOT project. At the same time, the tower can grow as in height, that is, fall lower on a very stable concrete base and without problems scale the power at times. And there is no windmill. This is its limit with the scope of the wind wheel of several hundred meters. At the same time, the weight of the windmill is several tens of times greater.

Oh well, that's why we cover 94% of the solar radiation falling into the ocean and distributed in the currents. And we can bring the power of HYPOT to the required values. And marine windmills have their own limit, depending on the huge size, and perceive only 6% of the solar radiation reflected from the surface of the ocean and distributed in the atmosphere. And do not forget that the density of water is 800 times higher than the density of air, that is, the energy losses in the twigs are simply not comparable to HYPOT.

HYPOT perceives the potential energy of the water hammer, which accumulates as the liquid moves in the collector and almost completely stops it in front of the neck of the tower. When a water hammer occurs in milliseconds, the speed increases according to Bernoulli's law and water gushes into an area of low pressure. Therefore, the process of wave, that is, the incoming kinetic energy is quantized. That is, the process is subject to quantum mechanics, and not just put a windmill in the wind.

HYPOT will completely abandon the construction of dams on rivers. They are no longer needed. Mankind has been using the power of falling water for centuries to obtain electrical energy. Hydroelectric power plants have been operating for decades and affect the climate. Apparently, this is why most people deny a fundamentally new source of energy from water rising up. The conversion of potential energy into kinetic energy occurs into a rollback Gravitational energy is accumulated in water, which is used in soliton therefore the HYPOT has following advantages:

1. First of all, it is the ability to scale HYPOT to the required consumption of electrical energy. Orbital Marine Power (OMP) limited to 2 MW, so to increase power, you need to keep afloat a larger number of boats at anchor. At the same time, they have all the disadvantages of keeping the boat in a stable position on the surface. This complicates operation. HPT has no such limitations.

2. Further, OMP can perceive only strong surface currents and will not be able to perceive bottom currents. In the event of a storm or other phenomena on the surface, strong excitement can immediately stop the operation of the power plant for an indefinite time.

3. OMP in full calm also stops working and can only work in constant tidal currents, that is, in certain places where it is

4. OMP propellers are absolutely not protected from collision with objects passing by it or uncontrolled ships. The areas of operation of such power plant will have to be limited by a barrier given the swing of anchors of several hundreds of meters.

5. HYPOT is hidden under water and on the surface only hatches the nozzle of the tower of a small cross-section. The blades of the vortex turbine are fully protected from environmental impact and the operation of the hydroelectric power plant does not affect the flora and fauna by the noise of the screws.

6. The presence on OPM of service mechanisms for lowering propellers, greatly affects the cost of manufacturing and maintenance, which is deprived of HYPOT.

7. At the same time, it should be noted that the epidural of the distribution of the flow power in the river varies in a large direction from the bottom to the surface, and in the sea - vice versa. This is a physical law and must be followed when designing power plants on the high seas.

In marine conditions, the underwater power plant should operate mainly on the bottom, in the lower reaches, although its design allows it to work in a suspended pontoon state, but in any case, its design should be simple and easily replaceable, which maximizes the profitability of electricity generation and reduces production costs.

6. Development

On the basis of fundamental and applied interdisciplinary research, this project considers the development of methodological, engineering and technological foundations for the creation of a new generation of environmentally friendly and cost-effective technologies and autonomous energy systems based on the use of kinetic and potential energy of bottom and surface currents resulting from changes in temperature and pressure at different depths of the World Ocean and continental rivers.

The project solves the problem of creating efficient energy technologies for autonomous decentralized power supply of offshore oil platforms, including on the Arctic shelf, using new generation underwater power plants and intelligent automated control systems.

The central problem requires consideration and solution of a number of subtasks:

Development of modern computational methods and calculation tools, digital design methods, materials and technologies for the creation of underwater hydroelectric power plants with vertical spiral blades of medium and high power with high hydrodynamic characteristics and structural strength for the conditions of real runoff of bottom and surface waters, as well as climatic conditions inherent in the northern territories.

Analysis, research and development of the theory of intelligent control of the underwater power grid of cyclone HPPs on the example of the use of a primary energy source with a "random" or "stable" nature of the arrival, its reliable forecasting for different time intervals and the development of software and technologies for its effective use.

7. Findings

The development of computers, in particular cluster technologies, allows the use of computational methods of hydrodynamics in the calculation of viscous currents in turbomachines. The introduction of numerical modeling in the process of development and research of the device allows you to reduce the cost of subsequent experimental refinement, and ideally abandon it.

This gives more freedom when solving problems of optimizing the geometry of the blade and other elements, external problems, without resorting to the formulation of the experiment.

To simulate the characteristics and calculation of spiral hydraulic turbines, methods of CFD analysis with large grids comparable in number of cells with calculations of non-sequencer processes, as well as the need for calculations on several low-detectable points to obtain the maximum efficiency mode and increase the energy eclipse of air ducts through four-blade horizontal-axial acceleration using active control systems, new profiles and geo-optimization are proposed blade metrics.

Preliminary results of mathematical tests showed significant results from the possible introduction of such power plants, which can be seen from the attached graph of the dependence of the hydraulic power capacity of the power plant on water consumption.

To study the effect of water hammers on renewable energy, it is proposed to create a pulsed shock wave generator that reproduces shocks close to real ones, and studies their effect on fragments of carbon fiber blades of spiral turbogenerators. Experimental studies of the influence of these effects on the blades are proposed.

As part of the task of digital design of elements and structures of a hydroelectric power plant, it is proposed to develop a design model of a blade system operating in real natural and climatic conditions, conduct CFD analysis using a high-performance cluster and build a 3D model of the blade that has better hydrodynamic performance and less weight compared to analogues.

The solution to the problem of creating a methodology for digital design of the conditions of the Far North (working under the ice) is interdisciplinary and complex: both known proven and tested methods from various branches of science and technology will be thoroughly studied and applied.

Scientifically based technical and technological solutions obtained during the work will be used to improve energy efficiency, efficiency, reliability, safety and technology in the North Sea.

The result of this approach will be the search for solutions for maximum autonomy of power plants without maintenance for a long time, respectively, the proposed systems will be more focused on self-healing, diagnostics and reconfiguration.

Analysis of the problem of building decentralized energy systems based on renewable hydropower sources using the theory of intelligent control.

Analysis and research of modern theoretical and applied issues of calculation, modeling and design of hydro turbine gravitational hydroelectric power plants for their manufacture using a new automated production technology.

Analysis of existing systems of active regulation of fluid flow in marine energy applications.

The analysis of the modern CAD world is adapted for end-to-end digital design of marine gravitational energy sources.

As part of the study of the existing scientific base, a method for forecasting underwater marine and channel river hydropower resources in different time intervals will be developed in order to configure the proposed energy device with the development of an interdepartmental approach to solving project problems.

Notes/acknowledgements/other statements

I want to express my deep gratitude to my daughter Helen and friends who took part in the request for a project to study a new renewable energy source.

Conflict of interest

The authors state that there is no conflict of interest.

Abbreviations and abbreviations

HYPOT	Hydropower Tower

Author details

George Mamulashvili
Highest Attestation Commission of the Russian Federation, Moscow, Russia

*Address all correspondence to: Gsmprado2018@gmail.com

References

[1] Petostatic tube. Copyright © 2021 Elsevier B.V. or its licensors or authors. ScienceDirect ® is a registered trademark of Elsevier B.V. [Internet] Available at: https://www.sciencedirect.com/topics/engineering/pitot-static-tube/

[2] Generating renewable power from water hammer 2 pressure surges A. Robertsa,∗ , B. Thomasa,, P. Sewella,, E. Hoarea, 3 4 aDepartment of Design and Engineering, Bournemouth University, Poole, Dorset, BH12 5 5BB [Internet] is available at: http://eprints.bournemouth.ac.uk/30975/1/renewable-1-BT.pdf;

[3] Hydraulic Pump Analysis and Improvement Using Computational Fluid Dynamics (CFD) by Piyush B. Shende Dr. S. K. Choudhary, A. P. Ninawe, IJIRST – International Journal of Innovative Research in Science and Technology| Volume 2 | Issue 03 | August 2015 ISSN (online): 2349-6010 All rights reserved www.ijirst.org 109 [Internet] Available at: http://www.ijirst.org/articles/IJIRSTV2I3043.pdf

[4] Orbital Marine Energy (Orkney Islands) plc. Orbital 02 2 MW Tidal turbine. EMEC Pier 5, Fall of War, Edei, Orkney Islands. A summary of the project [Internet] is available at: https://marine.gov.scot/sites/default/files/project_information_summary_4.pd f/ SR2-2000 Draft Information Document for a Maritime License.

[5] Water Hammer 1st Edition Practical Solutions 0.0 star rating Write a review Authors: Bruce Sharp David Sharp Hardcover ISBN: 9780340645970 eBook ISBN: 9780080543673 Imprint: Butterworth-Heinemann Published Date: 1st December 1995 Page Count: 192ISBN 978–0-340-64597-0 Language English Published 1995 Copyright Copyright © 1995 Elsevier Ltd. All rights reserved Butter Imprintworth-Hemann DOI Available at: https://www.elsevier.com/books/water-hammer/9780340645970

[6] Handbook of Shock Waves Front Matter Copyright PREFACE CONTRIBUTORS Volume Contents SBN 978-0-12-086430-0 Language English Published 2001 Copyright Copyright © 2001 Elsevier Inc. All rights reserved Imprint Academic Press. Available at: https://www.sciencedirect.com/book/9780120864300/handbook-of-shock-waves#book-info

[7] Arlan Skortegaña AlmeidaVladimir Caramori Borges de Sousa. An alternative method for measuring velocities in open channel flows: evaluating the performance of a Pitot tube compared to an acoustic meter. Available at: https://www.scielo.br/j/rbrh/a/TqFxzyvfnHQZMky9c6nRQsq/?lang=en#/

[8] Research and innovation of the European Green Deal (H2020-LC-2020). [Internet] Available at: https://waterquality.danube-region.eu/new-calls-are-available-at-euroaccess/.

[9] SeaGen-S 2MW - SIMEC Atlantis Energy [Internet] is available at: https://atlantisresourcesltd.com/wp/wp-content/uploads/2016/08/SeaGen-Brochure.pdf;

[10] US7832979B2.Vortex hydraulic turbine. United States. Inventor: Metin Ilbay Yaras; Mohammad Golriz. Application US11/736,766 event. [Internet] Available at: https://patents.google.com/patent/US7832979B2/en?oq=US+Patent+7832979

Section 3

Hydropower Managment Development

Chapter 4

Improved Memetic Algorithm for Economic Load Dispatch in a Large Hydropower Plant

Ling Shang, Xiaofei Li, Haifeng Shi, Feng Kong and Ying Wang

Abstract

This paper is intended to study the method of solving the economic load dispatch problem (ELDP) of hydropower plants via using memetic algorithm. Based on characteristics of economical operation of the hydropower plant, this paper proposes an improvement method of mutation operator and selection operator of memetic algorithm. Taking Three Gorges hydropower station in China as an example, the performance of memetic algorithm before and after improvement is tested separately. The test result shows that the average water consumption for simulation of the improved memetic algorithm is less than that for simulation of the standard memetic algorithm by 1.35%–16.19%. When the total load of the hydropower station is low (8GW-10GW), the water consumption for the improved memetic algorithm is less than that for the standard memetic algorithm by more than 10%. When the total load of the hydropower station is high (11GW-16GW), the water consumption for the improved memetic algorithm is less than that for the standard memetic algorithm by more than 1%. This shows that improvement of mutation operator and selection operator can improve the global and local optimization capacity of memetic algorithm a lot indeed. In addition, by comparing the optimization result of memetic algorithm with that of DP algorithm, it finds that the optimization result of improved memetic algorithm can reach the same precision of optimization result of DP algorithm. Therefore, using the improved memetic algorithm to solve the ELDP problem of large hydropower stations is practical and feasible. Since "curse of dimensionality" may occur frequently while using DP algorithm to solve the ELDP problem of large hydropower plants, as a new heuristic algorithm, memetic algorithm has obvious advantages in solving large-scale, complex, highly-dimensional and dynamic problems.

Keywords: economic load dispatch problem, Three Gorges hydropower station, memetic algorithm

1. Introduction

All the time, humans are inspired from the nature and discover many natural laws by observing and thinking natural phenomena. People have obtained abundant inspirations for solving various problems based on these natural laws and their own thoughts [1]. In 1975, Holland proposed a stochastic optimization algorithm by reference to the natural law of "survival of the fittest" in the biosphere and the algorithm was called genetic algorithm (GA) later [2]. Although the genetic algorithm has a strong global optimization capacity, it has many defects [3]. For example, the genetic algorithm searches the whole objective group but only converges to one optimal solution finally. To improve the addressing efficiency, the genetic algorithm does not search all populations and individuals, causing failure to explore the whole available space. Hence, diversity of populations is lost and finally GA cannot find out the real optimal solution correctly but can only find out the relatively optimal solution [4]. In 1989, Moscato proposed to combine global search based on population with local heuristic search based on individual, so as to prevent insufficient search scope of genetic algorithm and prematurity of the algorithm [5, 6]. This is the primary design concept of memetic algorithm (MA). At first, the memetic algorithm was regarded as improvement of genetic algorithm and therefore was called "hybrid genetic algorithm" [7, 8]. With continuous research, the memetic algorithm has been developed into a general evolutionary algorithm framework consisting of global search strategy and local search strategy [9, 10].

Within the framework, great improvement space exists in the memetic algorithm. For a specific problem, a memetic algorithm suitable for the problem can be constructed flexibly [11–13]. During construction of the memetic algorithm, on the basis that the global search strategy of genetic algorithm is reserved, the researcher generally proposes an improvement scheme of local zone search strategy based on the problem characteristics, so as to construct various memetic algorithms. For example, Yeh [14] developed a memetic algorithm by combining a genetic algorithm and the greedy heuristic using the pairwise exchange method and the insert method, to solve the flowshop scheduling problem. Boughaci et al. [15] proposed an improved memetic algorithm by using a stochastic local search (SLS) component combined with a specific crossover operator. The resulting algorithm is proved to be able to solve the optimal winner determination problem in combinatorial auctions. Zou [16] solved the traveling salesperson problem (TSP) by using an improved memetic algorithm. The algorithm applies multiple local search strategies and each search operator executes with a predefined probability to increase the diversity of the population, so as to ensure a higher search efficiency of the algorithm. Castro et al. [17] proposed a new memetic algorithm for solving the traveling salesperson problem (TSP) with hotel selection. Using tabu search algorithm and individual neighborhood information as the meta-heuristics search algorithm and genetic algorithm as the global search strategy may obtain several feasible schemes in one time. In fact, the memetic algorithm has achieved a very good application effect in solving the scheduling problem in material distributing and supply chain management [18–20].

In recent years, the memetic algorithm attracts more and more attention from researchers of other industries and the algorithm achieves considerable development with continuous efforts of many researchers [21–24]. Ammaruekarat and Meesad [25] proposed an improved multi-objective memetic algorithms (MOMAs) for solving the multi-objective decision problem. This paper proposes a new iterative search strategy—Chaos Search and uses it as the local search strategy of the memetic algorithm. Combining with Chaos theorem, the efficiency of solving the multi-objective decision problem will be improved a lot and good results are achieved. Özcan et al. [26] developed an Interleaved Constructive Memetic Algorithm (ICMA), and successfully applied ICMA for Timetabling problems with complicated and challenging structures. In addition, the memetic algorithm is also frequently applied to the neural network training algorithm. O'Hara and Bull [27] and Abbass [28] use the memetic algorithm to train the neutral network separately and believe that the effect of neuron network training with the memetic algorithm is better than that with the traditional method. Bonfim and Yamakami [29] not only use the memetic algorithm to train the neural network system, but also use it for Parallel Machine Scheduling, both which have achieved good effects.

Economical operation of hydropower plants is a very complex multi-objective optimization difficulty [30, 31]. Using traditional methods such as equal increment and dynamic programming for solving the problem cannot achieve perfect results [32, 33]. Especially for large hydropower stations, "curse of dimensionality" may occur frequently while using the dynamic programming algorithm for unit load dispatch, causing that load dispatch cannot meet the requirement of real-time control or the load cannot be dispatched at some times [34]. Three Gorges hydropower station in China is the hydropower plant with the largest installed capacity in the world at present; 32 units of seven types are installed; the capacity of single unit is 700,000 kW; and the total installed capacity is up to 22,400,000 kW. Although the generating capacity per unit is the same, the output character of unit is quite different due to different manufacturer of the unit [35]. Thus, using the simplified generating efficiency function to describe the generating character of all units is improper. Similarly, expressing the generating efficiency as the linear function of head will cause a big difference between the calculation result and actual operation condition [36].

Taking Three Gorges hydropower station in China (the largest hydropower station in the world) as an example, this paper studies the modeling method of economical operation model of Three Gorges hydropower station and proposes a method of using memetic algorithm to solve load dispatch and real-time scheduling of generating unit of Three Gorges hydropower station. On this basis, this paper refers to the optimization idea of differential evolution algorithm, further optimizes the solving process of ELDP problem and improves the memetic algorithm. The structure of residual parts of this paper is as follows: introduce the economical operation model of hydropower plant at first, introduce dynamic programming suitable for solving of the model, describe the standard memetic algorithm and its improvement method, use the memetic algorithm for solving of economical operation model of hydropower plant, compare the performance of memetic algorithm with that of dynamic programming, and discuss the possibility of application of memetic algorithm to the ELDP problem of hydropower plant. At last, this paper gives conclusions and looks forward to the application prospect of memetic algorithm.

2. Methodology

2.1 Formation of ELDP problem of hydropower plant

The ELDP problem of hydropower plant means that when the required load of system (the daily load chart is given generally in short-term economical operation) is determined, the consumed power discharge of the whole hydropower plant shall be the minimum, so as to obtain the maximum economic benefit of the hydropower plant [37]. For the economical operation model of hydropower plant, the minimum power discharge is the objective. The objective function and constraint conditions of the ELDP problem of hydropower plant are as follows:

Objective function:

$$\min Q = \sum_{i=1}^{n} q_i(N_i, H) \tag{1}$$

Load balancing constraint:

$$\min Q = \sum_{i=1}^{n} q_i(N_i, H) \tag{2}$$

Constraint of operation condition zone of unit:

$$0 \leq N_i \leq NH_i, \text{and } N_i \notin \left[\underline{N_{i,j}}, \overline{N_{i,j}}\right], j \in \Omega_i(H) \tag{3}$$

Where, Q is the total power discharge; n is the number of units; $q_i(\bullet)$ is the power discharge of unit i (m^3/s); N_i is the output of unit i (MW); H is the average head of time periods (m); Nd is the required load of power grid (MW); NH_i is the expected output of unit i (MW); $?_i(H)$ is the set of vibration zones of unit i; $\overline{N_{i,j}}$ and $\underline{N_{i,j}}$ are the upper and lower limits of vibration zone j of unit i at the given head H respectively (MW).

It should be noted that the unit of hydropower plant will break through the constraint of unit operation condition inevitably during operation and operate in the restricted operation zone or even in the forbidden operation zone. If the unit operates in the restricted operation zone for a long time, the flow passage components will be damaged; the output and efficiency will be reduced; and noise and strong vibration of unit will be caused. In severe cases, longitudinal cracks will occur on the dam, endangering safety of the hydropower station and surrounding areas. To meet actual demands, not only the restriction of expected output of unit but also the requirements of avoiding unit cavitation zone/vibration zone are considered in the constraint of unit operation condition zone in the above model. In the optimization algorithm design later in this paper, the study will apply the penalty function method to handle the output-flow relation curve of unit cavitation zone/vibration zone, so as to make the unit avoid the unsafe operation zone as far as possible during load dispatch.

2.2 Solution approaches

2.2.1 Dynamic programming (DP): A review

Dynamic programming method is a classic and mature optimization algorithm. It regards the ELDP problem of hydropower plant as a multi-stage decision problem and has no restrictions on that whether the unit model is the same. In addition, the number of operating units, combination of units, load and corresponding water consumption of each unit can be obtained as well as the globally optimal solution [38]. The dynamic programming algorithm is used to solve the ELDP problem of hydropower plant. Details of the variable definition, constraint handling and solving process are as follows:

2.2.1.1 Stage variable and state variable

The serial number i of generator unit is the stage variable while the cumulative output of i units ($\sum_{t=1}^{i} N_t$) is the state variable.

The step size of sate discretization is dN and the cumulative output is subject to state discretization as per the following formula:

$$Ns_{i,j} = \begin{cases} \min\left\{ j \bullet dN, \sum_{t=1}^{i} NY_t, Nd \right\}, & \textit{when } i \neq n \\ Nd, & \textit{when } i = n \end{cases} \tag{4}$$

Where,

$$j = 0 \sim \text{int}\left[\min\left\{\sum_{t=1}^{i} NY_t, ND\right\}/dN\right] + 1$$

$Ns_{i,j}$ is the state variable value at stage i and state j; NY_t is the installed capacity of unit t; and int[•] is the Gauss rounding function.

2.2.1.2 Constraint handling

While using the penalty function to handle the output condition constraint of unit, consider that the objective function of penalty quantity at stage I ($f_i(N_i, H)$) is as follows:

$$f_i(N_i, H) = q_i(N_i, H) + \Delta q_i + \Delta qp_i \tag{5}$$

$$\Delta q_i = \alpha_1 \bullet INF, \Delta qp_i = \alpha_2 \bullet INF \tag{6}$$

Where,

$$\alpha_1 = \begin{cases} 1 & N_i \in \left[\underline{N_{i,j}}, \overline{N_{i,j}}\right], \exists j \\ 0 & N_i \notin \left[\underline{N_{i,j}}, \overline{N_{i,j}}\right], \forall j \end{cases} \tag{7}$$

$$\alpha_2 = \begin{cases} 1 & N_i \in (-\infty, 0) \cup (NH_i, +\infty) \\ 0 & N_i \in [0, NH_i] \end{cases} \tag{8}$$

Where, Δq_i is the penalty term of constraint of operation condition zone and Δqp_i is the penalty term of constraint of output definition domain. α_1 and α_2 are the penalty coefficients and INF is the maximum.

2.2.1.3 State transition and state traversal

Using N_i as the decision variable, the state transition equation can be written as follows:

$$\sum_{t=1}^{i} N_t = \sum_{t=1}^{i-1} N_t + N_i \tag{9}$$

Recurrence equation:

$$Q_i^*\left(\sum_{t=1}^{i} N_t\right) = \min\left\{ f_i(N_i, N) + Q_{i-1}^*\left(\sum_{t=1}^{i-1} N_t\right)\right\} \tag{10}$$

Where, $Q_i^*\left(\sum_{t=1}^{i} N_t\right)$ is the optimal cumulative power discharge in the remaining period.

2.2.2 Memetic algorithm

Local search is the root cause that the memetic algorithm is better than the genetic algorithm. Through local search, the search depth of the algorithm for the solution space is increased a lot, further improving the solution quality. In this paper, the memetic algorithm using simplex optimization method for local search operation is called standard memetic algorithm. In the following, we will introduce how to use the memetic algorithm to solve the ELDP problem of hydropower plant and give emphasis on the improvement method of standard memetic algorithm for the ELDP problem of hydropower plant.

2.2.2.1 Standard memetic algorithm (SMA)

For the ELDP problem of hydropower station, the memetic algorithm applies integer encoding and establishes a corresponding relation between encoding and cumulative output value of unit. Details of encoding of memetic algorithm, initial population generation method, definition of fitness function and its main operators are as follows:

2.2.2.1.1 Gene encoding

The cumulative output of i units ($\sum_{t=1}^{i} N_t$) is defined as genes. Based on Formula (4) for discrete units with discrete step length defined as dN', the genes are encoded as $p_{k,i} = 0 \sim \left[\min\left\{\sum_{t=1}^{i} NY_t, Nd\right\}/dN'\right] + 1$ to represent the element sequence of $Ns_{i,j}$.

The cumulative output of i units is decoded as $Ns_{i,p_{k,i}}$ ($k = 1 \sim Pop, i = 1 \sim n, Pop$ stands for population and n stands for number of units).

2.2.2.1.2 Initial population generation

The linear constrained elimination method is used to generate the genes in reverse order under the conditions of load balance and output domain constraints.

When the cumulative output of i units is known as $\sum_{t=1}^{i} N_t$, then $p_{k,i} = int\left[\sum_{t=1}^{i} N_t/dN'\right]$.

The state transition equation (9) can be re-written as:

$$\sum_{t=1}^{i-1} N_t = \sum_{t=1}^{i} N_t - N_i \tag{11}$$

When the output feasible region constraint of unit $N_i \in [0, NY_i]$ is integrated into Formula (11), then:

$$\sum_{t=1}^{i-1} N_t \in \left[\sum_{t=1}^{i} N_t - NY_i, \sum_{t=1}^{i} N_t\right] \tag{12}$$

As the output is bound to be smaller than the installed capacity, i.e. $N_t \in [0, NY_t]$, to integrate it into Formula (11):

$$\sum_{t=1}^{i-1} N_t \in \left[0, \sum_{t=1}^{i-1} NY_t\right] \tag{13}$$

The common solution to Formulas (12) and (13) can satisfy the requirement for output domain, then:

$$\sum_{t=1}^{i-1} N_t \in \left[\max\left\{ \sum_{t=1}^{i} N_t - NY_i, 0 \right\}, \min\left\{ \sum_{t=1}^{i} N_t, \sum_{t=1}^{i-1} NY_t \right\} \right] \tag{14}$$

Making $\underline{Ntmp} = \max\left\{\sum_{t=1}^{i} N_t - NY_i, 0\right\}$, $\overline{Ntmp} = \min\left\{\sum_{t=1}^{i} N_t, \sum_{t=1}^{i-1} NY_t\right\}$, the generating approach for gene $p_{k,i-1}$ is expressed as:

$$p_{k,i-1} = int\left[\underline{Ntmp}/dN'\right] + int\left[Rnd \bullet \left(int\left[\overline{Ntmp}/dN'\right] - int\left[\underline{Ntmp}/dN'\right]\right)\right] \tag{15}$$

Where, Rnd indicates a random number evenly distributed in the internal [0,1].

Given load balance $\sum_{i=1}^{n} N_i = Nd$, the reverse recursion of Formulas (14) and (15) from the last gene is performed to obtain individuals that satisfy output domain and load balancing constraints.

2.2.2.1.3 Fitness function

According to the objective function, the fitness formula is constructed:

$$Fitness = \frac{INF}{\sum_{i=1}^{n} f_i\left(Ns_{i,p_{k,i}} - Ns_{i-1,p_{k,i-1}}, H\right)} \tag{16}$$

2.2.2.1.4 Crossover operator

For *Pop* individuals initially generated, select two individuals as per the preset crossover probability for crossover operation and generate a new generation of group (two new individuals).

$$X_1^{new} = \omega_1 \bullet X_1 + (1 - \omega_1) \bullet X_2 \tag{17}$$

$$X_2^{new} = \omega_2 \bullet X_2 + (1 - \omega_2) \bullet X_1 \tag{18}$$

X_1 and X_2 are two parent individuals selected from the population at random; X_1^{new} and X_2^{new} are new offspring individuals generated by crossover operation; and ω_1 and ω_2 are parameters selected from [0,1] at random.

2.2.2.1.5 Mutation operator

Among new individuals generated by crossover operation, select several individuals as per a certain mutation probability and conduct mutation operation as per the mutation operator in the following formula:

$$V'_{i,j} = \begin{cases} X_{i,j} + \left(b_{sup} - X_{i,j}\right)\left[r \bullet (1-t)^2\right], sign = 0 \\ X_{i,j} - \left(X_{i,j} - b_{inf}\right)\left[r \bullet (1-t)^2\right], sign = 1 \end{cases} \tag{19}$$

Where, $X_{i,j}$ is the component j of selected mutation individual X_i; $V'_{i,j}$ is the individual after mutation; *sign* is 0 or 1 at random; b_{sup} and b_{inf} are the upper and lower limits of parameters respectively; r is the random number from [0,1]; and t is the population evolution mark and $t = g_c/G_{\max}$, where g_c is the current evolution algebra of population while $G_{\max}$ is the maximum evolution algebra of population.

2.2.2.1.6 Selection operator

Select *Pop* excellent individuals from the current population, make them have the chance to be selected to the next iteration process, and abandon individuals with low fitness. The probability of each individual to be selected is in direct proportion to its fitness and the selection probability is as shown in Formula (20):

$$p_i' = \frac{\frac{1}{f_i}}{\sum \frac{1}{f_i}} \tag{20}$$

2.2.2.1.7 Local search operation

The simplex method is used for local search operation, namely, for the convex polyhedron consisting of $n + 1$ peaks ($X_1, X_2, \dots, X_n, X_{n+1}$) in n-dimensional space, calculate function values of $n + 1$ peaks and confirm the worst peak X_w, secondary bad peak X_s and optimal peak X_b and the centroid X_m of all points other than the worst peak in the simplex:

$$X_m = (X_1 + X_2 + \cdots + X_n + X_{n+1} - X_w)/n \tag{21}$$

Then, calculate the reflection point X_r passing X_m and X_w:

$$X_r = X_m + (X_m - X_w) \tag{22}$$

There are three possible conditions for the reflection point:

- If X_r is better than X_b, calculate the extension point X_c along the reflection direction, $X_c = X_m - \alpha \bullet (X_m - X_w)$, where, $\alpha > 1$ is the extension coefficient. If X_c is better than X_b, use X_c to replace X_w and generate a new simplex; or use X_r to replace X_w and generate a new simplex.

- If X_r is worse than X_b but not worse than X_s, use X_r to replace X_w and generate a new simplex.

- If X_r is worse than X_s, it indicates that X_r is too far and it shall be compressed along directions of X_r and X_m. Making X_h be the relatively optimal point between X_r and X_w, calculate the compression point X_c, $X_c = X_m - \beta \bullet (X_h - X_m)$, where, $0 < \beta < 1$ is the compression coefficient. If X_c is not worse than X_h, use X_c to replace X_w and generate a new simplex. Or conduct compression of simplex, namely, use X as the base point and cut the initial simplex into a half.

In all the above conditions, the new simplex must have a peak better than certain peak of the initial simplex; the simplex is subject to reflection, extension and compression through circulation; and the search process may converge to certain locally optimal solution or may be completed till meeting the termination condition.

2.2.2.2 The improvement to SMA (improved MA, IMA)

Differential Evolution Algorithm is also an effective technique to solve complex optimization problems, which is widely used in the fields of parameter optimization, neural network training, robot, energy and so on [39–42]. The Differential Evolution Algorithm in essence is a kind of greedy genetic algorithm based on real number encoding with the idea of guaranteeing optimality, which solves the optimization problems through the cooperation and competition among individuals in the population [43]. In this paper, the optimization idea of Differential Evolution Algorithm is used for reference to improve the mutation operators and selection operators of Standard Memetic Algorithm (SMA). The improvement methods for the mutation operators and selection operators of SMA are as follows:

2.2.2.2.1 Improvement to mutation operator

In differential evolution, the mutation operation uses the linear combination of multiple individuals in the parent population to generate new individuals, of which the most standard mutation component is the difference vector of the parent individual. For any target individual X_i in the parent population, the mutation individual V_i is generated according to the formula (23):

$$V_i = X_{r_1} + F(X_{r_2} - X_{r_3}), i = 1, 2, \cdots Pop \tag{23}$$

Where, $\{X_{r_1}, X_{r_2}, X_{r_3}\}$ are three different individuals randomly selected from the parent population, and $r_1 \neq r_2 \neq r_3 \neq i$, in which F is the zoom factor and the value range is [0,2], which is used to control the influence of the difference vector $(X_{r_2} - X_{r_3})$.

2.2.2.2.2 Improvement to selection operator

The "greedy" selection method [44] is adopted in the selection operation, and V_i is accepted by the population if and only if the fitness value of the new vector individual V_i is better than that of the target vector individual X_i. Otherwise, X_i will remain in the next generation population and continue to perform mutation and crossover operations as the target vector in the next iterative computation. For the minimization problem:

$$X_i^{t+1} = \begin{cases} V_i & f(V_i) < f(X_i^t) \\ X_i^t & Other \end{cases} \tag{24}$$

The selection operation is the one-to-one competition between the parent individuals and the newly generated candidate individuals to select the superior and eliminate the inferior, so that the offspring individuals are always superior to or equal to the parent individuals and thus the population can always evolve towards the optimal solution.

2.2.2.3 Workflow of memetic algorithm

Figure 1 shows the flow diagram of the Memetic Algorithm. The execution steps of the Memetic Algorithm are as follows: ① to determine the coding scheme of the problem and set the relevant parameters; ② to initialize the population; ③ to execute the crossover operator; ④ to use the local search algorithm to conduct neighborhood search for individuals, and update all individuals. ⑤ To execute the mutation operator to generate new individuals. ⑥ To use the local search algorithm to conduct neighborhood search for individuals again, and update all individuals. ⑦ To calculate the fitness value of all individuals in the population through the fitness function.⑧ To execute the selection operator for the population screening as per the natural law of "survival of the fittest" to abandon the individuals with poor fitness and retain the individuals with high fitness. ⑨ To determine whether the

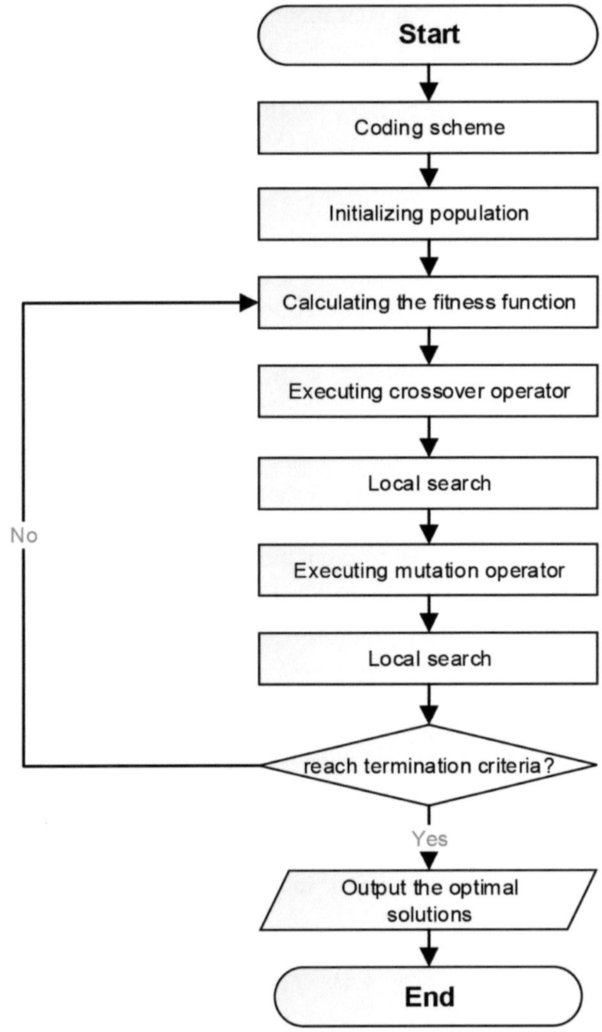

Figure 1.
Flow diagram of the memetic algorithm.

termination conditions are met. To determine whether the optimization criteria or the termination conditions of the algorithm have been met; if met, terminate the operation, otherwise continue to execute step ③.

3. Experiment, results and analysis

3.1 Experimental setup

The Three Gorges Hydropower Station is equipped with 14 generating units on the left bank and 12 on the right bank. Currently, 26 units of the power plants on the left and right banks have been automatically put into the power grid operation. The configuration of these units in the Three Gorges Hydropower Station is shown in **Figure 2**. Those 26 units can be classified into 5 categories: # 1 - # 3 and # 7 - # 9 are VGS; # 4 - # 6 and # 10 - # 14 are ALSTOM I; # 15 - # 18 are ORIENTAL I; # 19 - # 22 are ALSTOM II; and # 23 - # 26 are HARBIN. The output curves of the five types of units differ greatly. The unit output curves are shown in **Figure 3**.

Under the operating condition that the head of Three Gorges Hydropower Station is at 108 m, the unit load distribution performance of Memetic Algorithm before and after the improvement are tested in this paper. The overall water consumption of the hydropower station is mainly analyzed in the circumstance that the power grid load is increased step by step from 8GW to 16GW and the load distribution is carried out as per the unit load distribution results by Memetic Algorithm. Whether the algorithm performance is good or bad is determined by analyzing the water consumption under a given load. And it is believed that the smaller the total water consumption, the better the algorithm load distribution results, and the better the algorithm performance. In this analysis, 26 units that are automatically put into the power grid operation are taken as the research object in this paper, and the result of the DP algorithm is used as the benchmark for comparison.

The population size of the Memetic Algorithm is set to be 100, then the crossover probability $P_c = 0.6$, the mutation probability $P_m = 0.5$, the penalty factor $\lambda = 4000$, and the maximum generation $G_{\max} = 300$. Considering that the Memetic Algorithm is a stochastic optimization algorithm, the solution has a certain degree of uncertainty. In order to eliminate the influence of the randomness of initial solution on the calculation results, two kinds of Memetic Algorithms before and after the improvement are used in this paper to respectively carry out 10 operations for each load level, from which the best results are selected as the final optimal solution, and the average value of the operation results is used as the final result for analysis.

3.2 Results and analysis

For the nine load levels of 8GW, 9GW, 10GW, 11GW, 12GW, 13GW, 14GW, 15GW and 16GW, the load distribution schemes of the Memetic Algorithm before improvement (SMA) and of the Improved SMA (ISMA) are used respectively to calculate the total water consumption of 26 generating units of the Three Gorges Hydropower Station, which is shown in **Figure 4**.

It can be seen from **Figure 4** that with the increase of the total load of the hydropower station, the total water consumption of the hydropower station is always on the rise no matter whether the load distribution scheme of the Standard Memetic Algorithm (SMA) or of the improved Memetic Algorithm (ISMA) is adopted. However, on the whole, the average water consumption in the improved Memetic Algorithm is always lower than that of the standard Memetic Algorithm, which shows that the improved Memetic Algorithm reduces the water consumption rate of power generation and improves the utilization efficiency to water resources. **Figure 5** shows the reduction of the water consumption of the Three Gorges Hydropower Plant by ISMA compared with that by SMA.

It can be seen from **Figure 5** that the average water consumption for simulation of the improved Memetic Algorithm is less than that for simulation of the standard Memetic Algorithm by 1.35%–16.19%. The improved Memetic Algorithm saves more than 10% of the water consumption compared with the standard Memetic Algorithm when the total load of the power station is relatively low (8GW-10GW)

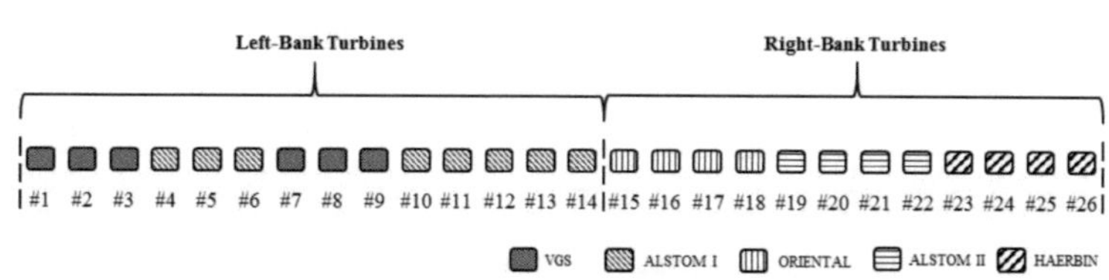

Figure 2.
Layout of generating units on the left and right banks of the Three Gorges Hydropower Plant.

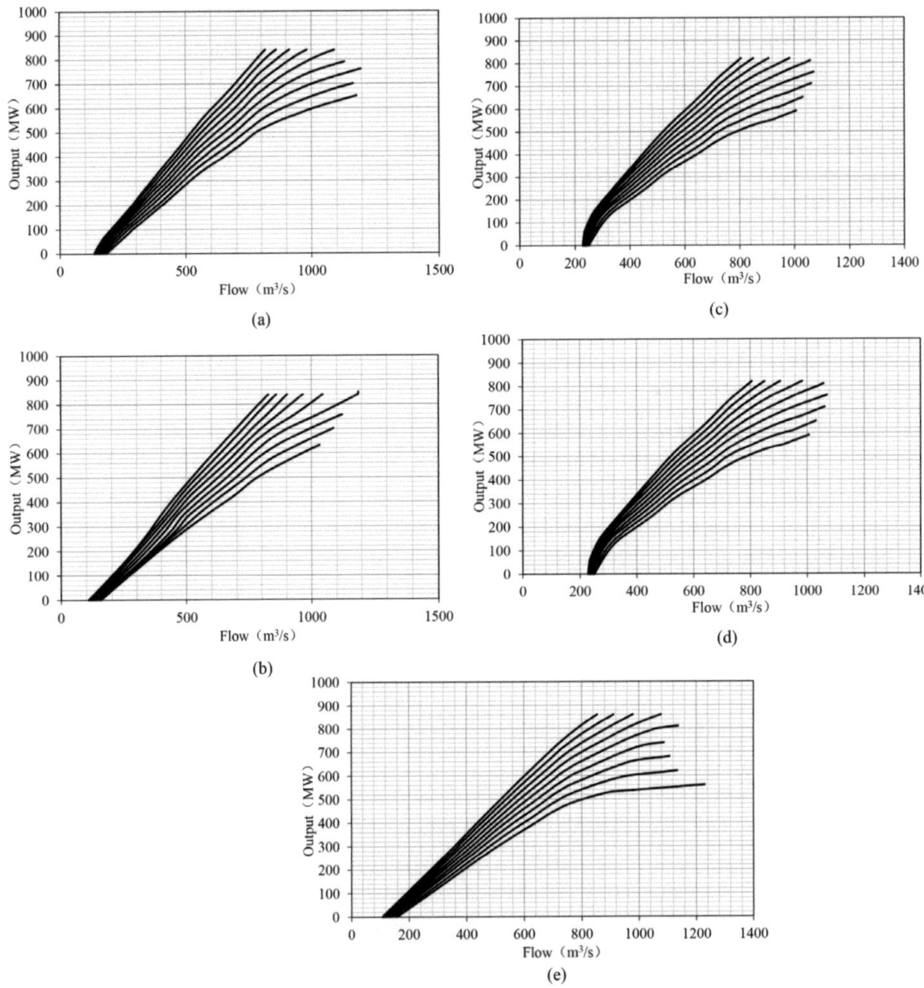

Figure 3.
Output curve graphs of the five types of units of the Three Gorges Hydropower Plant (head at 70–110 m, variation interval of water level at 5 m). (a) VGS. (b) ALSTOM I. (c) ORIENTAL. (d) ALSTOM II. (e) HAERBIN.

and also saves more than 1% of water consumption when the total load of the power station is relatively high (11GW-16GW). This shows that the Memetic Algorithm which improves the mutation operation and the selection operation enhances the global and local optimization capacities of the Memetic Algorithm.

In order to further compare the performances of the two algorithms, the evolutionary processes of the two algorithms at the total load of the Three Gorges Hydropower Station of 8GW, 10 GW, 12GW and 14 GW are analyzed in this paper, which is shown in **Figure 6**. As shown in **Figure 6**, the optimal water consumption obtained by the Standard Memetic Algorithm stimulation decreases in smaller range with the increase of the evolution algebra; the optimal water consumption obtained by the improved Memetic Algorithm stimulation decreases obviously with the increase of the evolution algebra, which shows that the improved Memetic Algorithm can perform an effective global search at the early stage of evolution and make the individuals in the population move closer to the globally optimal solution quickly, compared to which the Standard Memetic Algorithm has a weak global searching ability.

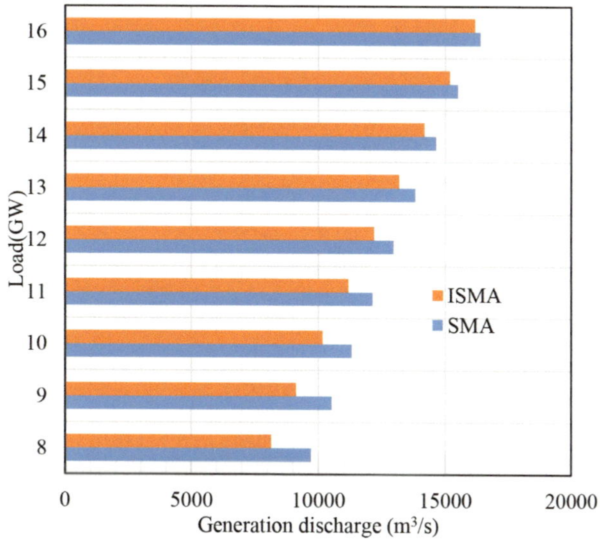

Figure 4.
Total water consumption of the Three Gorges Hydropower Station under the load distribution carried out respectively as per SMA and ISMA calculation schemes.

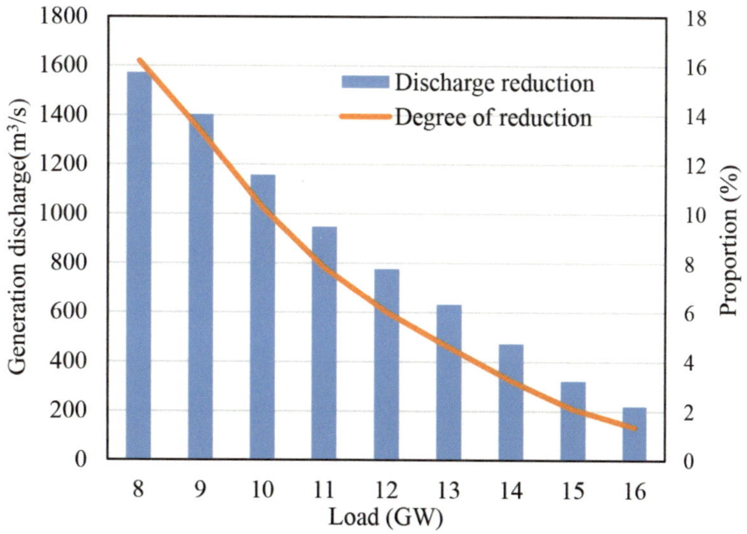

Figure 5.
Analysis of performance improvement degree of ISMA compared with that of SMA.

In this paper, the optimal simulation results of the two Memetic Algorithms are compared with the DP optimization results under the same load discrete interval. See **Figure 7** for the comparison results. As can be seen from **Figure 7**, the optimization result of the improved Memetic Algorithm is almost identical with the DP calculation accuracy, while the result of the Standard Memetic Algorithm stimulation is poorer, which shows that compared with the Standard Memetic Algorithm, the improved Memetic Algorithm can effectively search the globally optimal solution. The comparison results indicate that it is feasible to use the improved Memetic Algorithm to solve the problem of economic operation in a large hydropower station.

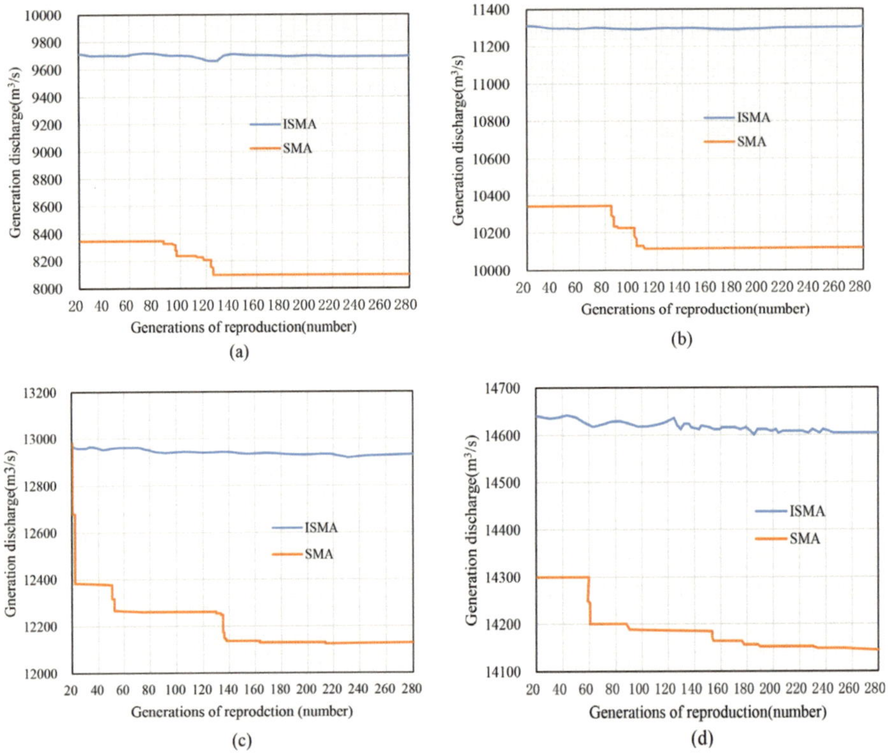

Figure 6.
Evolutionary processes of SMA and ISMA. (a) At the load of 8GW. (b) At the load of 10GW. (c) At the load of 12GW. (d) At the load of 14GW.

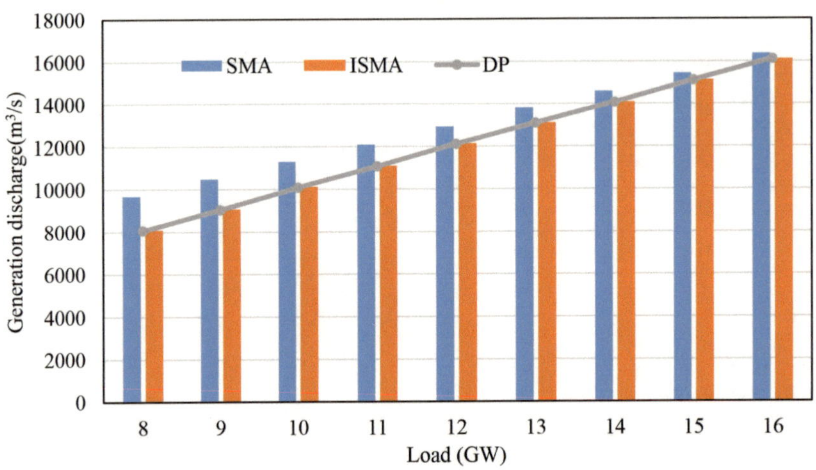

Figure 7.
Diagram of comparison between the optimal simulation results of the two memetic algorithms and the optimization results of DP algorithm.

4. Conclusion and discussion

Through the in-depth study of biology, scientists gradually find that individuals in the nature behave in a simple manner and with very limited ability, but when they work together, what they show is not a simple superposition of individual capabilities but very shocking and complex behavior characteristics. Inspired by this, scientists are paying more and more attention to the study of swarm intelligence algorithm, which also includes the study of the Memetic Algorithm. The combination of excellent global searching ability and fast local searching ability can produce enormous energy, therefore, the Memetic Algorithm has been paid more and more attention and has been recognized by more scientists. With the continuous in-depth study, Memetic Algorithm has not only improved the genetic algorithm but also has developed into a loose framework of the optimization algorithm.

In order to solve the problem of economic operation in hydropower plant, a new Memetic Algorithm is presented in this paper. Through the combination with the local search strategy, the Memetic Algorithm not only inherits the global optimization capacity of genetic algorithm itself, but also greatly improves the local searching ability of the algorithm by locally adjusting the new individuals generated by evolution. The framework and operational process similar with that of the genetic algorithm are used in Memetic Algorithm, but the Memetic Algorithm has an additional local search optimization process after crossover and mutation. Memetic Algorithm fully absorbs the advantages of genetic algorithm and local search algorithm and adopts the operational process of genetic algorithm, but after each crossover and mutation, local search is carried out, where the bad population will be removed early by optimizing the population distribution, thus reducing the iterations and speeding up the rate of convergence of the algorithm.

Taking China Three Gorges Hydropower Station, the largest hydropower station in the world, as an example, this paper studies the method of using Memetic Algorithm to solve the problem of economic operation in hydropower plant. The experiment result shows that it is feasible to use the Memetic Algorithm to solve the problem of economic operation in a large hydropower station. The experiment result also shows that the Memetic Algorithm improved by the idea of differential evolution demonstrates a better load distribution performance when compared with that before improvement. When the total load of the hydropower station is relatively low (8GW-10GW), the water consumption for the improved Memetic Algorithm is less than that for the standard Memetic Algorithm by more than 10%. When the total load of the hydropower station is relatively high (11GW-16GW), the water consumption for the improved Memetic Algorithm is also less than that for the standard Memetic Algorithm by more than 1%. This shows that the improvement of mutation operation and selection operation can greatly enhance the global and local optimization capacity of Memetic Algorithm.

As an intelligent algorithm of high efficiency, Memetic Algorithm has obvious advantages in solving large-scale, complex, high-dimensional and dynamic problems. However, the Memetic Algorithm needs a lot of improvements, for example, how to choose global search operators and local search operators, how to determine the control parameters such as population size, crossover probability and mutation probability, etc. All of these problems need to be further improved. In the future research work, the emphasis will be focused on the following aspects for in-depth study: ① How to further optimize the Memetic Algorithm framework to make the algorithm more flexible. ② To carry out in-depth study for the key control parameters to find out the setting rules of the control parameters. ③ To apply the optimized Memetic Algorithm to a wider field.

Acknowledgements

This work was supported by the natural science foundation of Nanjing vocational college of information technology under grant No.YK20190401; High level introduction of talent research start-up fund of Nanjing vocational college of information technology under grant No.YK20200501.

Author details

Ling Shang[1*], Xiaofei Li[2], Haifeng Shi[1], Feng Kong[1] and Ying Wang[1]

1 Nanjing Vocational College of Information Technology, Nanjing, Jiangsu Province, China

2 Beijing Sany Architectural Design Research Co., Ltd, Hunan, China

*Address all correspondence to: shangling@njcit.cn

References

[1] Grosan, C., Abraham, A. Hybrid Evolutionary Algorithms: Methodologies, Architectures, and Reviews. In: Abraham A., Grosan C., Ishibuchi H. (eds.) Hybrid Evolutionary Algorithms. Studies in Computational Intelligence, Springer, Berlin, Heidelberg. 2007, 75:1-17.

[2] Kumar, M., Husian, M., Upreti, N., Gupta, D. Genetic algorithm: Review and application. International Journal of Information Technology and Knowledge Management, 2010, 2(2): 451-454.

[3] Tan, K.C., Lee, T. H., Khor, E.F. Evolutionary algorithms for multi-objective optimization: Performance assessments and comparisons. Artificial Intelligence Review, 2002, 17 (4): 251-290.

[4] Sivaraj, R., Ravichandran, D. T. A review of selection methods in genetic algorithm. International Journal of Engineering Science and Technology, 2011, 3(5):3792-3797.

[5] Krasnogor, N. Memetic Algorithms. In: Rozenberg G., Bäck T., Kok J.N. (eds.) Handbook of Natural Computing. Springer, Berlin, Heidelberg, 2012, 905-935.

[6] Wang, Y., Hao, J.K., Glover, F., Lü, Z. A tabu search based memetic algorithm for the maximum diversity problem. Engineering Applications of Artificial Intelligence, 2014, 27:103-114.

[7] Arab, A., Alfi, A. An adaptive gradient descent-based local search in memetic algorithm applied to optimal controller design. Information Sciences, 2015, 299:117-142.

[8] Radcliffe, N.J., Surry, P.D. Formal memetic algorithms. In: Fogarty T.C. (eds.) Evolutionary Computing. AISB EC 1994. Lecture Notes in Computer Science, 1994 (865). Springer, Berlin, Heidelberg.

[9] Garzafabre, M., Kandathil, S.M., Handl, J., Knowles, J., Lovell, S.C. Generating, maintaining, and exploiting diversity in a Memetic Algorithm for protein structure prediction. 2016, 24 (4):577-607.

[10] Hu, Z., Bao, Y., Xiong, T. Comprehensive learning particle swarm optimization based memetic algorithm for model selection in short-term load forecasting using support vector regression. Applied Soft Computing, 2014, 25:15-25.

[11] Pishvaee, M.S., Farahani, R.Z., Dullaert, W. A memetic algorithm for bi-objective integrated forward/reverse logistics network design. Computers and Operations Research, 2010, 37 (6): 1100-1112

[12] Pan, Q.K., Wang, L., Sang, H.Y., Li, J.Q., Liu, M. A high performing Memetic Algorithm for the Flowshop scheduling problem with blocking. IEEE Transactions on Automation Science and Engineering, 2013, 10 (3):741-756.

[13] Acilar, A.M., Arslan, A. A novel approach for designing adaptive fuzzy classifiers based on the combination of an artificial immune network and a memetic algorithm. Information Sciences, 2014, 264: 158-181.

[14] Yeh, W. C. A memetic algorithm for the N/2/Flowshop Alpha F plus Beta C-Max scheduling problem. International Journal of Advanced Manufacturing Technology, 2002, 20 (6):464-473.

[15] Boughaci, Dalila, Benhamou, Belaïd, Drias, Habiba. A memetic algorithm for the optimal winner determination problem. Soft Computing, 2009, 13 (8-9):905-917.

[16] Zou, P., Zhou, Z., Chen, G. L., Yao, X. A novel memetic algorithm with random multi-local-search: a case study of TSP. Proceedings of the 2004 Congress on Evolutionary Computation, 2004, 2:2335-2340.

[17] Castro, M., Sörensen, K., Vansteenwegen, P., Goos, P. A memetic algorithm for the travelling salesperson problem with hotel selection. Computers and Operations Research, 2013, 40 (7):1716-1728.

[18] Divsalar, A., Vansteenwegen, P., Sörensen, K., Cattrysse, D. A memetic algorithm for the orienteering problem with hotel selection. European Journal of Operational Research, 2014, 237(1): 29-49.

[19] Hasani, A., Khosrojerdi, A. Robust global supply chain network design under disruption and uncertainty considering resilience strategies: A parallel memetic algorithm for a real-life case study. Transportation Research Part E: Logistics and Transportation Review, 2016, 87: 20-52.

[20] Bitar, A., Dauzère-Pérès, S., Yugma, C., Roussel, R. A memetic algorithm to solve an unrelated parallel machine scheduling problem with auxiliary resources in semiconductor manufacturing. Journal of Scheduling, 2016, 19 (4):367-376

[21] Deng, J., Wang, L. A competitive memetic algorithm for multi-objective distributed permutation flow shop scheduling problem. Swarm and Evolutionary Computation, 2017, 32: 121-131.

[22] Nalepa, J., Kawulok, M. Adaptive memetic algorithm enhanced with data geometry analysis to select training data for SVMs. Neurocomputing, 2016, 185: 113-132.

[23] Xue, X., Wang, Y., Ren, A. Optimizing ontology alignment through Memetic Algorithm based on Partial Reference Alignment. Expert Systems with Applications, 2014, 41(7): 3213-3222.

[24] Nekkaa, M., Boughaci, D. A memetic algorithm with support vector machine for feature selection and classification. Memetic Computing, 2015, 7 (1):59-73

[25] Ammaruekarat, P., Meesad, P. A chaos search for multi-objective memetic algorithm. Proceedings of International Conference on Information and Electronics Engineering (ICIEE 2011). 2011, 6: 140-144.

[26] Özcan, E., Parkes, A. J., Alkan, A. The Interleaved Constructive Memetic Algorithm and its application to timetabling. Computers and Operations Research, 2012, 39(10): 2310-2322.

[27] O'Hara, T., Bull, L. A memetic accuracy-based neural learning classifier system. IEEE Congress on Evolutionary Computation, 2005, 3:2040-2045.

[28] Abbass, H. A. A Memetic Pareto Evolutionary approach to artificial neural networks. Australian Joint Conference on Artificial Intelligence, 2001 (AI):1-12.

[29] Bonfim, T.R., Yamakami, A. Neural network applied to the coevolution of the memetic algorithm for solving the makespan minimization problem in parallel machine scheduling. Brazilian Symposium on Neural Networks, 2003, 44 (6):197.

[30] Shang, Y, Lu, S., Gong, J., Liu, R., Li, X., Fan, Q. Improved genetic algorithm for economic load dispatch in hydropower plants and comprehensive performance comparison with dynamic programming method. Journal of Hydrology, 2017, 554(C): 306-316.

[31] Fang, N., Zhou, J., Zhang, R., Liu, Y., Zhang, Y. A hybrid of real coded

genetic algorithm and artificial fish swarm algorithm for short-term optimal hydrothermal scheduling. International Journal of Electrical Power and Energy Systems, 2014, 62 (11):617-629.

[32] Li, C. L., Zhou, J. Z., Ouyang, S., Ding, X. L., Chen, L. Improved decomposition–coordination and discrete differential dynamic programming for optimization of large-scale hydropower system. Energy Conversion and Management. 2014, 84: 363–373.

[33] He, D., Wang, F. Mao, Z. A hybrid genetic algorithm approach based on differential evolution for economic dispatch with valve-point effect. International Journal of Electrical Power and Energy Systems. 2008, 30(1):31-39.

[34] Shang, Y., Lu, S, Ye, Y, Liu, R, Shang, L., Liu, C, Meng, X., Li, X., Fan, Q. China'energy-water nexus: Hydropower generation potential of joint operation of the Three Gorges and Qingjiang cascade reservoirs. Energy, 2018, 142:14-32.

[35] Li, X., Li, T., Wei, J., Wang, G., Yeh, W. Hydro unit commitment via mixed integer linear programming: A case study of the Three Gorges project, China. IEEE Transactions on Power Systems. 2014, 29(3):1232-1241.

[36] Séguin, S., Côté, P. Self-scheduling short-term unit commitment and loading problem. IEEE Transactions on Power Systems, 2016, 31(1):133-142.

[37] Suman, M., Rao, M Venu Gopala., Hanumaiah, A., Rajesh, K. Solution of economic load dispatch problem in power system using Lambda iteration and back propagation neural network methods. International Journal on Electrical Engineering and Informatics, 2016, 8(2):347-355.

[38] Liang, Z.X., Glover, J.D. A zoom feature for a dynamic programming solution to economic dispatch including transmission losses. IEEE Transactions on Power Systems, 1992, 7 (2):544-550.

[39] Sun, J., Zhang, Q., Tsang, E. P.K. DE/EDA: A new evolutionary algorithm for global optimization. Information Sciences, 2005, 169(3-4):249-262.

[40] Wang, L., Zeng, Y., Chen, T. Back propagation neural network with adaptive differential evolution algorithm for time series forecasting. Expert Systems with Applications, 42 (2): 855-863.

[41] Yildiz, A. R. A new hybrid differential evolution algorithm for the selection of optimal machining parameters in milling operations. Applied Soft Computing, 13(3): 1561-1566.

[42] Abido, M.A. A novel multiobjective evolutionary algorithm for environmental/economic power dispatch. Electric Power Systems Research, 65(1):71-81.

[43] Das, S., Suganthan, P.N. Differential evolution: A survey of the state-of-the-art. IEEE Transactions on Evolutionary Computation, 2011, 15 (1):4-31.

[44] Merz, P., Freisleben, B. Fitness landscapes, Memetic algorithms, and Greedy operators for graph bipartitioning. Evolutionary Computation, 2000, 8(1):61-91.

Section 4

Ecological Protection and Sustainability

Chapter 5

Innovative Projects and Technology Implementation in the Hydropower Sector

Emanuele Quaranta

Abstract

In this chapter, some innovative case studies in the hydropower sector are discussed, highlighting how novel technologies and operational practices can make it more efficient, sustainable and cost-effective. Some practices to reduce hydropeaking effects, improving fish habitat, and turbines with higher survival rate, allowing to bring fish survival >98%, are discussed. The retrofitting of non-powered barriers can help to minimize the environmental impacts, reducing costs by more than 20%. New turbines are described focusing on their advantages with respect to standard ones, in particular, water wheels in irrigation canals to promote the valorization of watermills and old weirs, the very low head (VLH) turbine in navigation locks (reducing overall cost by more than 20%), the vortex turbine, and the Deriaz turbine with adjustable runner blades to improve the efficiency curve, especially at part load. Digitalization can help in preventing damages and failures increasing the overall efficiency and energy generation by more than 1%.

Keywords: aqueduct, Deriaz, digitalization, fish, hydropeaking, retrofitting, turbine, sustainability, VLH, water wheel

1. Introduction

Hydropower is the largest renewable energy source used worldwide, with 1308 GW of global installed capacity in 2019. The benefits of large hydropower plants (>10 MW) with a reservoir are related to the multipurpose use of the reservoir, e.g. energy generation, job opportunities, better water management, storage capacity, and stabilization of the electric grid thanks to the flexible operation, while small hydropower (SH, <10 MW) typically contributes to local development, decentralized energy generation and market opportunities in remote areas [1, 7].

The hydropower sector is undergoing a technological evolution, due to the new needs it has to deal with. Hydropower is required to be more flexible, as the integration of wind and solar energy sources in the electric grid (characterized by an intermittent and highly variable production) are increasing the variability of the electricity market [1]. Furthermore, hydropower is required to be more environment friendly, as the interruption of the longitudinal continuity of a river and the related fragmentation generated by a dam generate severe environmental impacts that should be properly addressed [2–4]. Hydropower is also required to become cheaper, with a lower investment cost, so that the reuse of existing structures and the repowering of non-powered dams are perceived as a suitable option to reduce hydropower cost while avoiding additional river fragmentation [5].

Therefore, several emerging technologies and best practices are under development [6, 7] aimed at increasing hydropower flexibility, cost-effectiveness and sustainability, minimizing environmental impacts, and providing sustenance and electrification to rural areas [8]. More flexible turbines are being developed to cope with the always more frequent grid instabilities and load variations [9], pumped storage hydropower plants to allow storage capacity and flexibility [10], digitalization to prevent failures and to optimize the operation [6, 7], new low head hydropower converters to be used in irrigation canals and at low head existing barriers [11, 12], and more fish-friendly solutions to reduce impacts on fish [13].

In this chapter, some innovative case studies from Italy, USA, Belgium and Switzerland are discussed to show recent innovations in the hydropower sector. In particular, the discussed case studies are related to the following topics:

- Hydropeaking reduction
- Ecologically improved turbines
    a. Reaction turbines
    b. The vortex turbine
- Reuse of existing barriers:
    a. Waterways and basins
    b. Aqueducts
    c. Overflow from dams
    d. Weirs in irrigation canals
    e. Navigation locks
- New turbines: the Deriaz turbine
- Digitalization

These case studies are described with the aim of showing how the implementation of novel methodologies and technologies can help in reducing costs and impacts while increasing efficiency and energy generation. Proper references are included for the readers interested in knowing more about the technical details of these technologies, while here we will mostly focus on their benefits and effects.

2. Hydropeaking reduction

Hydropeaking refers to frequent, rapid, and short-term fluctuations in water flow and water levels downstream and upstream of hydropower stations. Such fluctuations are becoming increasingly common worldwide due to the variable electricity market and are known to have far-reaching effects on riverine vegetation

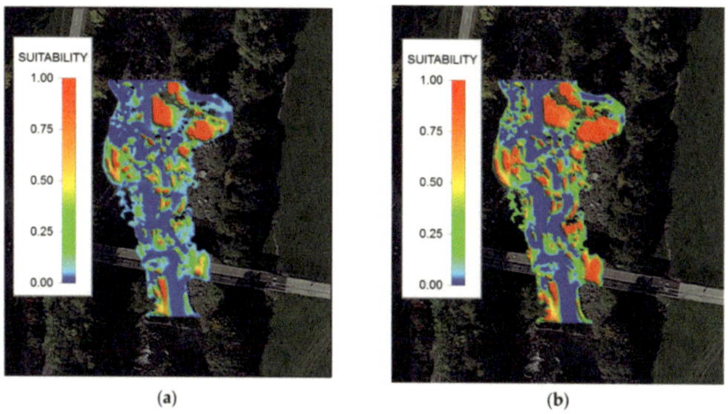

Figure 1.
(a) Suitability distribution for adult Salmo marmoratus and (b) young S. marmoratus. Q = 4.0 m^3/s [7].

and fish communities. The modified hydrology caused by hydropeaking has no natural correspondence in freshwater systems, and few species can adaptat [14]. Hydropeaking can be compared to artificial floods that drastically worsen the quality of the river environment [15].

In this context, the Sant Antonio hydroelectric plant was built in 1952 on the Talavera River in Bolzano, Italy. To mitigate the hydropeaking, the construction of two large demodulation reservoirs was chosen to store the water released from the turbines. The use of reservoirs is indeed considered one of the most effective methods to reduce hydropeaking [16]. The reservoirs gradually fill up when the turbines are operating and slowly empty when the plant works at part load.

Downstream of the plant the ratio between the maximum and minimum flow rate has been reduced from 1:15 to a value of 1:4. The ecological effects of the demodulation reservoir were estimated by examining water depth and flow velocity [17], depending on fish preferences and focusing on *Salmo marmoratus* and *Thymallus thymallus* (**Figure 1**).

The weighted usable habitat area (WUA) [18] was calculated to show how the underground demodulation reservoir improves the habitat availability both in the conditions of minimum release (WUA_{min}, occurring when the demodulation reservoir releases water) and in the maximum release conditions (WUA_{max}, when the peak flow rates are flattened).

The analysis of the expected effects on an annual basis shows for adult *Thymallus thymallus* an increase of 67.5% in WUA_{min} and a reduction of 1.2% in WUA_{max}, and for young *Thymallus thymallus* an increase of 23.4% in WUA_{min} and 6.3% in WUA_{max}. For adult *S. marmoratus,* it was estimated an increase of 14.6% in WUA_{min} and 2.9% in WUA_{max}, and for young *S. marmoratus* an increase of 3.4% in WUA_{min} and 7.5% in WUA_{max}.

3. Ecologically improved turbines

3.1 Reaction turbines

Ecologically improved turbines have gained attention in the past two decades and are designed with a strong focus on reducing the hydraulic stressors leading to mortal injury of migrating fish. A review work [13] on the first generation of "fish-friendly" turbines described the conceptual development and implementation of two relevant technologies designed for better fish passage conditions, namely the minimum gap runner (MGR) and the Alden turbine. Such technologies are continuously improving to give rise to a second generation of "fish-friendly" turbines that yield greater improvements in fish survival than the first generation (reducing blade-strike effects) and that accommodate a larger biodiversity of fish present in the migratory corridor. The pressure-related effects are important when there is a greater biodiversity since most migratory fish are prone to mortality due to barotrauma effects caused by the rapid decompression.

The Alden turbine has three blades wrapped around the shaft and it is the evolution of the Francis turbine. The Alden turbine rotates slower compared to conventional Francis turbines. Presently, the unit is only applicable for small-scale hydropower. Instead, the recent development of the Kaplan turbine is the minimum gap runner (MGR), the DIVE, and the very low head (VLH) turbine. Studies with such turbines report improved survival rates compared to the conventional design. The design of the minimum gap runner (MGR) reduces the gaps between the adjustable runner blade and the hub as well as between the blades and the discharge ring. In the DIVE turbines, the double regulation is provided by the variable rotational speed instead of the adjustable runner blade pitch. Such an approach allows maintaining the blades always in their maximum opening position reducing the strike probability with fish and avoiding dangerous gaps between the blades and other parts of the machine. The very low head (VLH) is adapted to sites below 5 m head, with fixed runner blades and adjustable rotational speed.

The U.S. Army Corps of Engineers (USACE) Ice Harbor Lock and Dam (**Figure 2**), located in Washington State, is one such facility focused on improvement under this point of view. The dam houses six large vertical Kaplan turbines. Units 1–3 produce 107 MW at 27.1 m of net head and were commissioned in 1961. Units 4–6 produce 130 MW.

The turbine runners and associated water passageway modifications were investigated with computational fluid dynamics to assess various fish passage criteria, including reducing shear stresses and a target minimum nadir pressure within the water passageway of atmospheric pressure, 101 kPa [19].

During the test procedure, 1068 treatment fish were released and 1030 were recaptured. At the peak efficiency condition, the 48-h survival estimates, excluding predation, were 98.16 ± 0.84%. A total of 15 of the recaptured treatment fish (1.5%) underwent visible injuries, in particular bruising to the head and body (0.7%) and eye damage (0.5%). Four fish (0.4%) were decapitated. Survival at 48 h at the peak efficiency was 2.2–3.3% higher for the new Unit 2 than at the existing adjustable blade runner tested in 2007 (Unit 3). About 1.5% of fish were injured at Unit 2, while 3.8% at Unit 3. About 0.4% of fish were decapitated at Unit 2, less than at Unit

Figure 2.
Ice Harbor Lock and Dam, located on the Snake River in southeast Washington State, USA [7].

3 (1.2%). It was also found that the target minimum nadir pressure of 101 kPa and the blade strike reduction were satisfied.

3.2 The vortex turbine

Hydraulic turbines with a free surface operation are generally considered fish-friendly, due to their large flow passages, no high-pressure gradient and no pressurized flows, and low rotational speeds [11, 29]. This is the case of the gravitational vortex turbine [7, 11].

The vortex turbine can operate within a head ranging from 1 to 4.5 m and with flow rates from 0.7 to 9 m^3/s. The current section describes the rehabilitation of a vortex turbine installed on the Ayung River in Bali. The head is 1.85 m and the flow is 1.5 m^3/s. The previous turbine was limited to 5 kW due to technical problems with the generator. The rotor often got blocked by debris, and, after a flood, the drivetrain underwent irreparable damages. The new runner has a diameter of 1.2 m and the concrete basin diameter is 3.9 m, where there is the vortical flow that drives the runner. The rotational speed is 96 rpm, with a generated power of 13 kW and thus with a global efficiency of 55.8%. Numerical simulations predicted good ecological behavior in relation to fish passages. The equivalent maximum acclimation pressure was P_a = 17.19 psa, while the minimum exposure pressure was P_e = −1.1 × 104 Pa (relative) = 13.1 psa, leading to P_e/P_a = 76%, above the threshold limit (60%). Furthermore, the maximum pressure drop rate is below 116 psi/s, which is below 500 psi/s, the threshold limit. There were also additional benefits of using this turbine, as the smaller dimensions and a lightweight turbine (rotor and drive train weight is 550 kg), and the relatively low cost (0.07 EUR/kWh including maintenance). Obviously, the civil structure of the concrete circular basin has to be considered, and future studies should aim at minimizing its dimensions.

4. Reuse of existing barriers

The powering of non-powered dams–NPDs—and the exploitation of existing barriers is one of the developing hydropower practices aimed at avoiding new interruptions of the longitudinal river connectivity [5]. For example, in USA there are 2500 dams that provide 78 gigawatts (GW) of conventional hydropower and 22 GW of pumped storage hydropower, but the United States has more than 80,000 NPDs, providing a variety of services ranging from water supply to inland navigation. Powering of these dams can add 12 GW, 8 GW of these from 100 dams [20]. Instead, a feasibility study conducted in the Piedmont region of North Carolina, cataloging over 1000 non-Federal dams with hydraulic heads ranging from 4.6 m to 10.7 and power capacity <300 kW, showed that most of the dams were not financially convenient for hydropower applications, although some low head dams may be exploited for hydropower applications if adequate funding opportunities were provided as for wind and solar markets [21]. In Europe, the main advantage of this approach is that most of the infrastructures are already in place and the requirements in the capital are between 30 and 50% of that for mini-hydro stations constructed from scratch. According to the European Environment Agency (EEA), there are currently approximately 7000 large dams in Europe and thousands of additional smaller dams. Thus, it is expected that a power potential exists in European NPDs. In South Africa, the potential of NPDs is estimated at 250 MW [22]. It is estimated that in Europe there are 65,000 small barriers, for example historic weirs and mill sites [23].

4.1 Waterways and basins

In this section, two historic channels, that provided water for irrigation to Modena and Reggio Emilia fed by a reservoir of 800,000 m^3, are examined (**Figure 3**). These channels are no longer feedable due to the lowering of the river bed. They are located in San Michele dei Mucchietti on the Secchia River, Sassuolo (MO, Italy), and Castellarano (RE, Italy). The water supplies are managed by legally different subjects with leading management in charge of Consorzio di Bonifica Emilia Centrale. The central part of the dam is made of a concrete body with a surface spillway and two bottom sluice gates; the right and left shoulders are made of earth.

Although the structure was not conceived to host a hydropower plant (HPP), a hydropower plant was successively designed to use the water-level difference generated by the weir without water resource subtraction to the riverbed. The main characteristics of the plant are average flow rate: 10.54 m^3/s, maximum flow rate: 26 m^3/s, gross head: 9.66 m, and nominal concession power: 998.20 kW.

Two vertical axis Kaplan turbines were installed to provide a maximum power of 1207 and 603 kW, connected to two synchronous generators with nominal power equal to 1476 and 800 kVA.

The plant was built by reusing part of the existing infrastructures, in particular:

1. Double connection pipes between the lateral basin and the river, which were originally oversized and used to allow the water to get back into the river from the lateral reservoir. This was done by separating one of the two pipes to separate the flows. 300,000 € were saved thanks to this design choice, because the bank-cutting operation of the lateral reservoir, and the restoration of the waterproof structure was avoided. The inlet front was covered by a concrete sill with the function of both stopping the hydropower intake to ensure the minimum level necessary for the irrigation and protecting from the silting of the intake.

2. The existing accesses were maintained, avoiding the construction of expensive new roads and trails.

3. The embankment of the lateral reservoir was maintained, saving 200,000 € for the construction of protection screens, huge stone boulders, walls higher than the two-hundred-year period flood level.

Therefore, 277 €/kW was the total saving that, compared to an average installation cost of 1300–8000 USD$/kW [24], is a cost reduction of 4.2–25%.

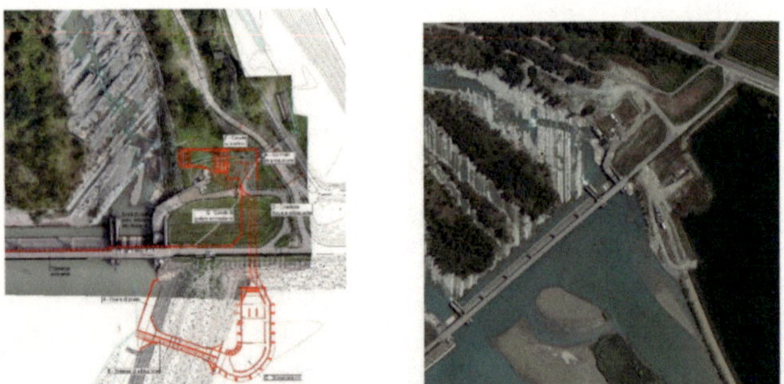

(a) (b)

Figure 3.
(a) Layout of the plant as in the design; (b) aerial photo of the site as built [7].

DOI: http://dx.doi.org/10.5772/intechopen.100492

4.2 Aqueducts

In existing water supply and irrigation networks, there is a hidden hydropower potential that can be exploited. For example, 4 bars are usually required in drinking water networks and 3–10 bars in irrigation plants equipped with sprinklers. If the hydrostatic pressure is higher than the above-mentioned requested operational pressure, the excess pressure can be exploited for hydropower production.

There are several benefits of implementing hydropower systems in existing pipelines: (i) head, pressure, and flow rate are continuously monitored, thus also the power generated; (ii) the placement of turbine can extend the life of nearby pressure reduction valves (PRV) as it reduces the workload of the PRV; and (iii) existing infrastructures are used; thus, some plants may be exempted from licensing, or the cost for permitting may be relatively small, and environmental impacts are reduced. However, as the water demand is not constant, the flow rate in pipelines varies substantially (±50%). Therefore, it is important to understand how turbines behave when subjected to these variations [25].

The Solcano HPP is located in the province of Pescara (Abruzzo region, central Italy) in the municipality of San Valentino in Abruzzo Citeriore. The nominal diameter 300 mm PN40 (40 bar) steel pipe feeds several municipalities with a maximum flow rate Q_{max} = 0.185 m^3/s.

The flow rate in the network varies substantially [26]. A Francis turbine was chosen and designed at Q = 0.125 m^3/s and N = 3000 rpm as design speed [27], but the wide head range made necessary a variable speed system. The installed power was 150 kW. This choice was justified by the fact that a faster synchronous speed was not possible, while a lower synchronous speed was not suitable for Francis turbines. Furthermore, multi-stage Francis turbines increase the complexity of the system and reduce reliability.

A fixed speed Francis solution would exhibit a very narrow operating range with respect to site operating range. Therefore, the variable speed solution, in this case, was selected, varying between 2500 and 4000 rpm.

4.3 Overflow from dams

In reservoir hydropower plants, the discharge necessary to preserve the river ecosystem and that has to be continuosly released downstream, called environmental flow, could be used to produce green energy without any other type of pollution. A case of energy recovery from environmental flow was realized in a mini-hydropower plant located in the North of Italy, in the alpine area of Predazzo close to Bellamonte, in the downstream part of Forte Buso dam. The main features of the Forte Buso plant are dam height 105 m, maximum retention height 99 m, basin storage volume 29.4 Mm^3, total head in maximum basin level 114 m (total head for the turbine set at level 0), and total head in minimum basin level 51 m.

The catchment hydrological basin is 60 km^2. Along the downstream Travignolo river, the following values of environmental flow must be released:

- December–March (winter): 392 l/s.
- April–July (spring) and October–November (autumn): 549 l/s.
- August–September (summer): 470.5 l/s.

The mini-hydropower uses the environmental flow. A permanent magnet generator (PMG) was chosen to face with the large head variation, and the turbine rotational speed varied from 500 to 600 rpm. The Pelton turbine is with a vertical axis and an installed power capacity of 580 kW, with six jets. The PMG designed power is 600 kW and 50 Hz. The plant includes a bypass system that works also in case of electricity failure, to always ensure the correct release of environmental flow. The annual energy generation is 3100 MWh, with an overall efficiency of 92.7%, higher than that of a traditional system (84–86%).

The overall cost was 3,000,000 €, of which 2,086,000 € was the cost of the execution of the plant. Therefore, the unitary cost was 5000 €/kW of installed power. The payback time is 6 years for the return of the investment. If two turbines would have been used, the cost would have been increased by 60%, while management and maintenance charges almost doubled. The payback time would have been increased to 9 years.

4.4 Weirs in irrigation canals and in old mills

In rivers, irrigation canals, and at old mill sites, there is a hydropower potential with very low head differences below 2.5 m. The power mostly ranges from 5 to 100 kW. Much of this potential is unused since modern hydropower technology is not cost-effective for such a very low head/high flow rate situation [28]. It is estimated that in Europe there are 65,000 unused historic hydropower sites, out of which 27,000 are old water mills that could be repowered by using water wheels [23].

Water wheels have been recently discovered to be efficient and cost-effective hydropower converters in this context, so that, in the last decade, the horizontal axis water wheel has been again reintroduced in the market for electricity generation. This was due to the hydraulic efficiencies of more than 70%, coupled with the low costs compared with other low head turbines [11] and high ecological behavior in relation to downstream migrating fish [29].

Horizontal axis wheels can be classified into gravity type and stream type. Gravity wheels mainly use the potential energy of water, that is, the water weight [12], and can be classified into overshot [30], breastshot [31], and undershot wheels [28] depending on head and flow rate, while stream water wheels use the kinetic energy of a water stream [32].

In this section, a breastshot water wheel realized in North Italy is presented. Breastshot water wheels are generally used below 4 m head and flows per meter width typically below 800 l/s. The water inflow is near the rotation axis.

Figure 4 depicts the water wheel installed in an irrigation canal in North Italy. The head difference is 1.85 m with a flow rate of 1.0 m^3/s. The wheel is made of

Figure 4.
Water wheel installed in an irrigation canal in North Italy.

COR-TEN steel. The wheel diameter is 3.6 m, for a wheel width of 1.35 m and 30 blades, supported by a shroud in the canal bed 2 m × 6 m. The global efficiency of the wheel and the power take-off (generator and gearbox) is estimated to be 0.67. The electric generator is a synchronous one, with 4 poles and 95.5 Nm of torque. Its efficiency is 0.95, as also the efficiency of the gearbox and of the inverter. The wheel weight is 51 kN, in agreement with the equation proposed in [33].

4.5 Navigation locks

Navigation locks present a perfect example of existing facilities where hydro turbines, instead of the gates, could regulate the flow, producing energy otherwise wasted. The very low head (VLH) machine is a promising technology in this context [7]. In this section, two VLH installations are described.

The first plant is located in the Canda locality (Rovigo, Italy) on the Canal Bianco, and it was commissioned at the end of 2016. The canal height is 7 m, canal width 9.5 m, gross head 3 m, and the mean annual flow 16.5 m^3/s. The second plant is located in Bussari (Rovigo, Italy), again on the Canal Bianco, and it was commissioned in the first half of 2017. The canal height is 6.3 m, canal width 8 m, gross head 2.56 m, and mean annual flow 25 m^3/s.

In the Canda power plant, two VLH 3150 (runner diameter of 3150 mm) were installed with a total power of 2 × 256 kW achieving an annual production of 2,888,000 kWh. In Bussari, one VLH 5000 (runner diameter of 5000 mm) was installed with a total power of 481 kW and annual production of 2,751,000 kWh.

Comparing this solution similar projects with Kaplan turbines, the civil costs were reduced by 80%, while the design of an *ad hoc* steel structure instead of a standard one lead to an increase of 40% of mechanical structures. Therefore, the total cost of the Canda and Bussari projects was 3950 and 3650 €/kW, respectively, lower than the typical cost of similar plants where civil works are needed (5000 €/kW). This means that the cost reduction ranged between 20 and 30%.

5. New turbines: the Deriaz turbine

Nowadays, climate changes, variable flow rates, unpredictable floods, and the frequent instability of the electric grid require new hydropower technologies to provide greater flexibility over an extended range of hydraulic conditions. Improved flexibility means maintaining an optimal efficiency while reducing flow instabilities, especially at off-design conditions, and efficiently respond to grid requirements and instabilities. This is provided by control techniques (1), new and more geometrically-adjustable turbines (2), better governors (3), and integration with other energy technologies (e.g. photovoltaic panels), additional reservoirs, and batteries (4) [6].

With regard to point (2), the Deriaz turbine, as optimization of the Francis turbine, is gathering a lot of attention. The Deriaz turbomachine (DT), presented by engineer Paul Deriaz, was the first diagonal flow pump-turbine to be designed, in 1926, but during the nineteenth century, it has been almost forgotten. The Deriaz turbine exhibits a lower specific speed than a Kaplan turbine, and it works more efficiently at part loads and at higher available head than a Kaplan turbine. Differently from Francis turbines, the Deriaz is provided with adjustable runner blades, providing higher flexibility at variable flow rates and with higher efficiency at part loads. The maximum efficiency is around 90%, and it can be kept constant from 30 to 120% of the design flow rate. The Deriaz turbine is able to work in pumping mode [34]. Kuromatagawa II power station adopted a vertical shaft Deriaz

(a) (b)

Figure 5.
(a) Deriaz turbine (photo courtesy of Mhylab); (b) Deriaz power plant in Italy (photo courtesy of Artingegneria).

pump-turbine in 1963 to deal with a large head variation and to attain high efficiency at part load. The rotational speed of the Deriaz turbine was 333 rpm during pumping mode and 300 rpm in turbine mode from an effective head of 78 m to the minimum operational head of 39 m with a maximum flow rate of 28 m^3/s. The ratio of unit speeds at maximum efficiency during turbine and pump operation is about 1.1, the same as for the Francis-type pump-turbine [34]. Another interesting example is a mini-hydropower plant built in Italy, with the following characteristics: maximum flow rate: 800 l/s, head: 32.5 m, power: 220 kW, 1000 rpm. The cost was 800 k€ (350 k€ of electro-mech. equipment) (**Figure 5**).

Also, the X-Blade turbine is an evolution of the Francis turbine. It can operate in a larger range of flow and heads before inter-blade vortices, inlet cavitation, or draft tube pressure pulsations occur or become critical. The skewed outlet geometry with the relatively small outlet diameter at the crown contributes to the typical low draft tube pressure pulsation level. The X-Blade turbine generates a smaller and less intensive vortex in the draft tube center than other turbine types, requiring a smaller amount of natural air admission through the runner cone.

6. Digitalization

The collection and processing of real-world data to adjust the actual working conditions of hydropower turbines can provide advanced grid supporting services without compromising reliability and safety. Apart from the improvement of predictive maintenance allowing for the prolongation of the lifetime, reduction of the outage time, and addressing cyber-security risks, rehabilitation and digitalization involve increasing the overall efficiency and, thus, the produced energy [6].

Recent studies have shown the interest in applying data-driven methods to the data collected during the hydropower plant operation or reduced scale model tests to predict fatigue and condition monitoring, as in [35], cavitation erosion in [36], and performances, as presented by [37].

Further information on the meaning and activity involved in the digitalization concept is well described in [6]. It is estimated that a total of 42 TWh could be added to present hydropower energy production by implementing hydropower digitalization. Such an increase could lead to annual operational savings of 5 $ billion and a significant reduction of greenhouse gas emissions. The increase of

42 TWh/y corresponds to 1.3% of the actual global hydro-generation. This is in line with a recent publication, [7] where it was calculated an additional 0.5% in one case, and 1.2% in a second case, of energy generation of two Italian hydropower plants by implementing the digitalization. The cases reported by Hydrogrid reached an efficiency increase between 0.4% and 1% [38].

Furthermore, digitalization will enable to drastically reduce the response time of hydro-units, especially those of reversible pump-turbines. In [39], it is presented the case study of Z'Mutt, a pumped storage hydropower plant equipped with a 5-MW reversible pump-turbine with variable speed technology using a Full-Size Frequency Converter (FSFC). It is showed that by leveraging numerical simulations and scale model test results during transient operations, and implementing an optimization algorithm to select the best operating sequences, the response time of the turbine can be improved. Digitalization will also allow to assess the economic impact of offering additional reserve flexibility, and to prevent failures and damages with the implementation of HPP digital twins. The cost of a predictive system for one unit (development and implementation at HPP) is about 200,000 EUR [40]. An additional increase and optimization could be achieved with the use of software that uses genetic algorithms such as EASY [41]. To improve further the flexibility of hydropower plants, a number of researchers [42] investigated their stability properties by means of transfer functions representing the dynamic behavior of the reservoir, penstock, surge tank, hydro-turbine, and the generator. A novel approach was developed to establish the dynamic model of the hydro-turbine governing system in the transient process.

HydEA is a platform to analyze the behavior of the plant and that elaborates a reference model of the performance characteristics of the generation units. This allows to detect in real time the deviations from the expected values, finding eventual damages. It is also possible to recognize, through the recalculation of the models at fixed intervals, very slow decay of the system performance. The additional algorithm allows to increase the overall plant efficiency by improving the load on the operating turbines. For example, the production of an Italian plant increased by 1.2% on an annual basis.

The Hydro-Clone is a Real-Time Simulation Monitoring System made of a numerical copy of the hydropower plant that reproduces its real-time dynamic behavior, using the boundary conditions measured *in situ* as input [43]. This system allows to continuously diagnose the health of a plant by numerical cloning the major hydraulic and electrical components of the plant, using the SIMSEN software. The comparison between the simulated and measured quantities enables to understand the health state and behavior of the system. The Hydro-Clone system has been operating since 2014 at the La Bâtiaz power plant (200 MW).

7. Discussion and conclusions

The case studies here collected show engineering insights on new technologies and more sustainable methodologies. These case studies confirm the fact that hydropower is a sector in continuous development. New technologies and methodologies are being implemented to improve flexibility and efficiency and to reduce environmental impacts. **Table 1** summarizes some key results.

New ecologically improved turbines are under development to reduce fish mortality and improve habitat, but more studies should be devoted to the better understanding of the interaction between fish and hydraulic structures. The Sant' Antonio hydroelectric plant uses two large underground demodulation reservoirs to reduce the effects of hydropeaking downstream of the plant, but more work is needed to better determine habitat preferences of some fish species [17].

Case study	Type of improvement	Improvement value
Hydropeaking reduction in Sant'Antonio plant	Fish habitat (WUA)	+67.5%
Ecologically improved turbine at Ice Harbor Lock and Dam	Increase in fish survival rate	>98%
Vortex turbine for low heads	Smaller runner dimensions and fish-friendly behavior	56% efficiency and pressure drop rate below 116 psi/s < 500 psi/s
Reuse of existing structures in San Michele dei Mucchietti on the Secchia River	Cost reduction	500,000 €, or 277 €/kW of cost reduction
Hydropower in aqueducts	Cost reduction and no environmental impacts	150 kW Francis turbine
Hydropower on environmental flow structures	Cost reduction	Payback time reduction from 9 to 6 years
VLH turbines in navigation locks	Cost reduction	Overall cost reduction up to 30%
Digitalization	Preventing failures and increasing efficiency	Less maintenance and increased efficiency +1%
New turbines: Deriaz turbines and water wheels	Flatter efficiency curve, and valorization of watermills	Higher weighted efficiency due to the double regulation, and energy at low head sites

Table 1.
Key improvements from the case studies.

The hydropower energy recovery from existing hydraulic structures is also an emerging trend. In San Michele dei Mucchietti locality, on the Secchia River, the use of existing structures has allowed saving 500,000 €, corresponding to a cost reduction of 277 €/kW. The energy generation from the ecological flow was also described, with a Pelton turbine with a global efficiency of 92.7%. If this facility would have been built with two standard turbines, the cost would have been increased by 60%. Energy recovery in aqueducts is also a sector in rapid development, and in this chapter, a 150-kW Francis turbine with variable speed was described [44].

The VLH turbine was implemented in a navigation canal in Italy, leading to a total cost between 3950 €/kW and 3650 €/kW, while similar plants in which civil works were needed had a total cost of 5000 €/kW. Compared to Kaplan turbines, the VLH turbine also shows a better ecological behavior, but it exhibits a lower efficiency and can only be applied at heads below 5 m.

Digitalization is an emerging trend, especially when hydropower plants have to be modernized (almost one half of the hydropower fleet was built more than 40 years ago). New tools are under development, allowing to improve annual generation spill reduction, and preventing damages and failures.

Acknowledgements

Thanks to Eisackwerk, Voith Hydro, Ada Francesconi, Carmine Fioravante of Artingegneria, and Alberto Bullani of Mhylab for the photo courtesy.

DOI: http://dx.doi.org/10.5772/intechopen.100492

Author details

Emanuele Quaranta
European Commission Joint Research Center (JRC), Ispra, Italy

*Address all correspondence to: emanuele.quaranta@ec.europa.eu; quarantaemanuele@yahoo.it

References

[1] Farahmand H, Jaehnert S, Aigner T, Huertas-Hernando D. Nordic hydropower flexibility and transmission expansion to support integration of North European wind power. Wind Energy. 2015;**18**(6):1075-1103

[2] Frey GW, Linke DM. Hydropower as a renewable and sustainable energy resource meeting global energy challenges in a reasonable way. Energy Policy. 2002;**30**(14):1261-1265

[3] Moran EF, Lopez MC, Moore N, Müller N, Hyndman DW. Sustainable hydropower in the 21st century. Proceedings of the National Academy of Sciences. 2018;**115**(47):11891-11898

[4] Quaranta E, Katopodis C, Revelli R, Comoglio C. Turbulent flow field comparison and related suitability for fish passage of a standard and a simplified low-gradient vertical slot fishway. River Research and Applications. 2017;**33**(8):1295-1305

[5] Hwee HH, Lu SF PHC, Fook PW. Converting dam: From water supply to hydropower generation. Journal of Engineering and Technological Advances. 2021;**2**(1):37-56

[6] Kougias I, Aggidis G, Avellan F, Deniz S, Lundin U, Moro A, et al. Analysis of emerging technologies in the hydropower sector. Renewable and Sustainable Energy Reviews. 2019;**113**:109257

[7] Quaranta E, Bonjean M, Cuvato D, Nicolet C, Dreyer M, Gaspoz A. Hydropower case study collection: Innovative low head and ecologically improved turbines, hydropower in existing infrastructures, hydropeaking reduction, digitalization and governing systems. Sustainability. 2020;**12**:8873

[8] Wang C. A Guide for Local Benefit Sharing in Hydropower Projects. Social Development Working Papers. Washington: The World Bank Social Development; 2012

[9] Trivedi C, Agnalt E, Dahlhaug OG. Investigations of unsteady pressure loading in a Francis turbine during variable-speed operation. Renewable Energy. 2017;**113**:397-410

[10] Rehman S, Al-Hadhrami LM, Alam MM. Pumped hydro energy storage system: A technological review. Renewable and Sustainable Energy Reviews. 2015;**44**:586-598

[11] Bozhinova S, Hecht V, Kisliakov D, Müller G, Schneider S. Hydropower converters with head differences below 2.5 m. Proceedings of the Institution of Civil Engineers-Energy. 2013;**166**(3):107-119

[12] Quaranta E, Revelli R. Gravity water wheels as a micro hydropower energy source: A review based on historic data, design methods, efficiencies and modern optimizations. Renewable and Sustainable Energy Reviews. 2018;**97**:414-427

[13] Hogan TW, Cada GF, Amaral SV. The status of environmentally enhanced hydropower turbines. Fisheries. 2014;**39**(4):164-172

[14] Bejarano MD, Jansson R, Nilsson C. The effects of hydropeaking on riverine plants: A review. Biological Reviews. 2018;**93**(1):658-673

[15] Bruder A, Tonolla D, Schweizer SP, Vollenweider S, Langhans SD, Wüest A. A conceptual framework for hydropeaking mitigation. Science of the Total Environment. 2016;**568**:1204-1212

[16] Heller P, Bollaert E, Schleiss AJ. Optimal reservoir use of a multipurpose run-of-river powerplant for

hydropeaking mitigation. In: Hydro2005, Villach, Austria-Policy into Practice. International Water Power and Dam Construction. 2005. pp. 1-8

[17] Negro G, Fenoglio S, Quaranta E, Comoglio C, Garzia I, Vezza P. Habitat preferences of Italian freshwater fish: A systematic review of data availability for applications of the MesoHABSIM model. Frontiers in Environmental Science. 2021;**305**

[18] Allen L, Conder T, Annear C. Test of weighted usable area estimates derived from a phabsim model for instream flow studies on trout streams. North American Journal of Fisheries Management. 1987;7:339-350

[19] Trumbo BA, Ahmann ML, Renholds JF, Brown RS, Colotelo AH, Deng ZD. Improving hydroturbine pressures to enhance salmon passage survival and recovery. Reviews in Fish Biology and Fisheries. 2014;**24**:955-965

[20] Hadjerioua B, Wei Y, Kao SC. An Assessment of Energy Potential at Non-Powered Dams in the United States. Oak Ridge, Tennessee: Oak Ridge National Laboratory; 2012

[21] Sandt CJ, Doyle MW. The hydrologic and economic feasibility of micro hydropower upfitting and integration of existing low-head dams in the United States. Energy Policy. 2013;**63**:261-271

[22] Patsialis T, Kougias I, Kazakis N, Theodossiou N, Droege P. Supporting renewables' penetration in remote areas through the transformation of non-powered dams. Energies. 2016;**9**(12):1054

[23] Punys P, Kvaraciejus A, Dumbrauskas A, Šilinis L, Popa B. An assessment of micro-hydropower potential at historic watermill, weir, and non-powered dam sites in selected EU countries. Renewable Energy. 2019;**133**:1108-1123

[24] Gielen D. Renewable energy technologies: Cost analysis series. Hydropower. 2012;**1**

[25] Sari MA, Badruzzaman M, Cherchi C, Swindle M, Ajami N, Jacangelo JG. Recent innovations and trends in in-conduit hydropower technologies and their applications in water distribution systems. Environmental Management. 2018;**228**:416-428

[26] Penninger, G, Benigni, H. Numerical simulation and design of spherical valves for modern pump storage power plants. In: Proceedings of the 14th International Seminar on Hydropower Plants; 297-305. Technische Universität Wien. 22-24 November 2006; Vienna, Austria, Institute for Waterpower and Pumps

[27] Quaranta E. Optimal rotational speed of Kaplan and Francis turbines with focus on low-head hydropower applications and dataset collection. Journal of Hydraulic Engineering. 2019;**145**:04019043

[28] Quaranta E, Müller G. Sagebien and Zuppinger water wheels for very low head hydropower applications. Journal of Hydraulic Research. 2018;**56**(4): 526-536

[29] Quaranta E, Wolter C. Sustainability assessment of hydropower water wheels with downstream migrating fish and blade strike modelling. Sustainable Energy Technologies and Assessments. 2021;**43**:100943

[30] Quaranta E, Revelli R. Performance optimization of overshot water wheels at high rotational speeds for hydropower applications. Journal of Hydraulic Engineering. 2020;**146**(9):06020011

[31] Quaranta E, Revelli R. Optimization of breastshot water wheels performance using different inflow configurations. Renewable Energy. 2016;**97**:243-251

[32] Quaranta E. Stream water wheels as renewable energy supply in flowing water: Theoretical considerations, performance assessment and design recommendations. Energy for Sustainable Development. 2018;**45**:96-109

[33] Quaranta E. Estimation of the permanent weight load of water wheels for civil engineering and hydropower applications and dataset collection. Sustainable Energy Technologies and Assessments. 2020;**40**:100776

[34] Morabito A, Silva GDO, Hendrick P. Deriaz pump-turbine for pumped hydro energy storage and micro applications. Journal of Energy Storage. 2019;**24**: 100788

[35] Åsnes A, Willersru A, Kretz F, Imsland L. Predictive maintenance and life cycle estimation for hydro power plants with real-time analytics. In: HYDRO 2018, Gdansk, Poland: Hydropower and Dams International

[36] Gregg SW, JPH S, Van Bossuyt DL. Machine learning: A tool for predicting cavitation erosion rates on turbine runners. Hydro Reviews. 2017;**36**(3)

[37] Bozic I, Jovanović R. Prediction of double-regulated hydraulic turbine on-cam energy characteristics by artificial neural networks approach. FME Transactions. 2016;**44**(2):125-132

[38] Madisetti K. Digital transformation in the hydro sector. In: BHA Annual Conference 2019, Hydrogrid, Glasgow: British Hydropower Association

[39] XFLEX HYDRO. Latest News on the Demonstrations. 2021. Available from: https://xflexhydro.net

[40] Georgievskaia E. Hydraulic turbines lifetime in terms of fracture mechanics. Engineering Failure Analysis. 2019;**105**: 1296-1305

[41] Kontoleontos E, Weissenberger S. Annual energy production maximization for tidal power plants with evolutionary algorithms. International Journal of Fluid Machinery and Systems. 2017;**10**(3):264-273

[42] Zhang J, Chen D, Zhang H, Xu B, Li H, Aggidis GA, et al. Fast–slow dynamic behaviors of a hydraulic generating system with multi-timescales. Journal of Vibration and Control. 2019;**25**:2863-2874

[43] Nicolet C, Béguin A, Bollaert E, Boulicaut B, Gros G. Real-time simulation monitoring system for hydro plant transient surveys. International Journal on Hydropower and Dams. 2015;**22**:62-69

[44] Mitrovic D, Chacón MC, García AM, Morillo JG, Diaz JAR, Ramos HM, et al. Multi-country scale assessment of available energy recovery potential using micro-hydropower in drinking, pressurised irrigation and wastewater networks, covering part of the eu. Water. 2021;**13**(7):899

Chapter 6

Hydropower and Sustainability

Hemlal Bhattarai

Abstract

Renewable energy sources are gaining momentum in power sector mainly to address the impacts of climate change as well as the risks associated with usage of fossil fuels or nuclear energy sources. Hydropower is one of the most promising renewable energy source-based power plant that hold significant shares globally. But there are series of risks associated with hydropower project when we talk about sustainability and needs are felt to critically understand the pertaining risks as well as protocols or measures to quantify the risks. Such measure will prove to be crucial in underlining the strategic measures from planning, construction and operation phases of hydropower keeping on account of its sustainability.

Keywords: hydropower, renewable energy, sustainability, protocols, risks

1. Introduction

Power sectors today are facing stiff challenges in terms of its growing roles in contributions towards socioeconomic development. Electrical energy is one major components of the contributors that drives economic activities with stiff increases in its demand. On other hand there is pressure of climate friendly adoption through the adoption of the principles of green economy whereby the need for greenhouse gas emissions needs to be reduced [1].

But power sectors have been facing immense pressure of its dependency for fossil fuels usage for the electric power generations. With the threat of climate change and the commitment to combat climate change there is considerable move to harness electrical energy from renewable energy which is a clean source of energy. Amongst the renewable sources, hydropower holds considerable shares of electrical power generation due to its stable high power generation capabilities.

2. Renewable energy sources for electricity generation

The initiative for renewable energy sources has been considerably gaining momentum in recent decades. Prime reasons for this are due to its positive impacts of being environment friendly as well as its inexhaustible properties. Statistics maintained by IRENA shows that globally there was addition renewable energy capacity by 260 GW in 2020 despite a year hard hit by COVID-19 pandemic. Major reflection of this growth attributes due to fall in global fossil fuel additions [2].

The (**Figure 1**) below clearly highlighted the growth of renewable energy where major shares are of hydropower followed by wind, solar, bioenergy and geothermal. Just in 2019 there seems a considerable growth in solar power followed by wind, and other sources for electricity generations [3].

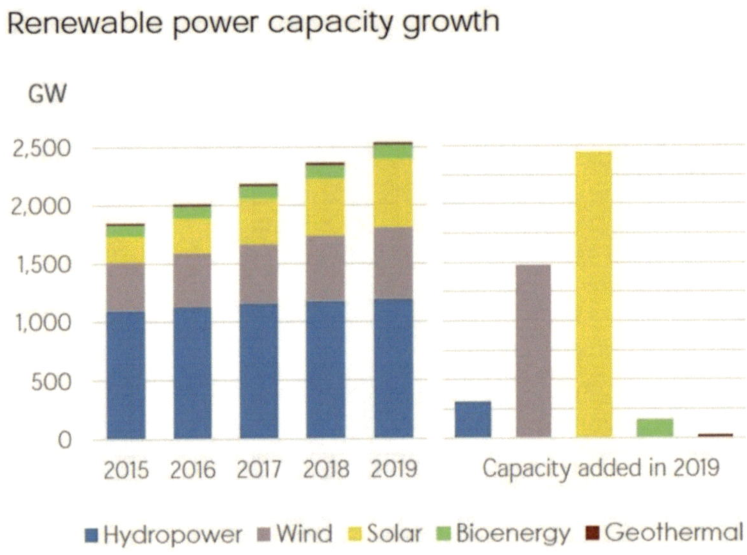

Figure 1.
Growth in renewable electric generation [2].

The share of installed capacity of hydropower in 2018 is at 1292 GW and there is a growth seen in the sectors of renewable energy (**Figure 2**). Such growth is due to be higher reliability of this source for renewable power generation along with its limited adverse impacts to the environment. As a result, there are indications that hydropower is gaining its potential across globe and the countries that can have feasibility of generating electric power from hydro. The trust based on its reliability assurance along with it being the source of renewable energy, penetration of hydropower in power sectors are substantial as well as growing.

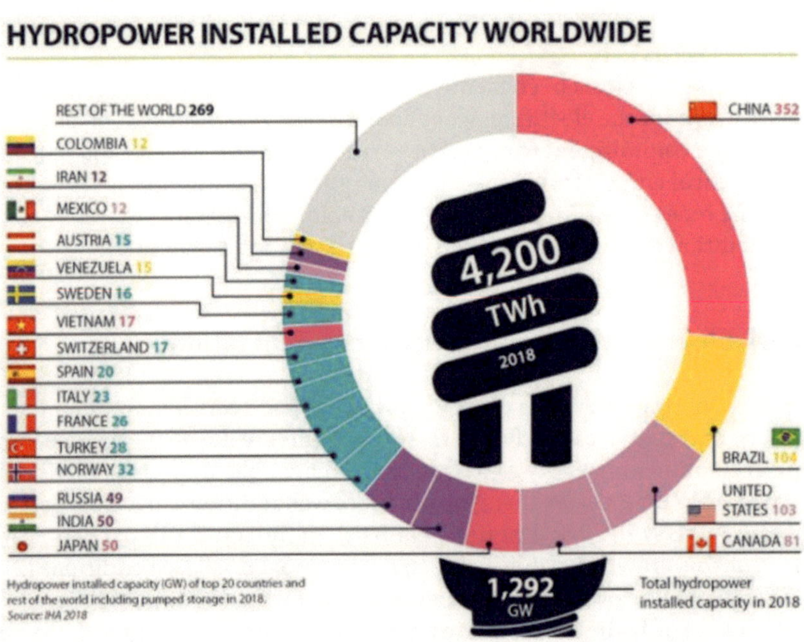

Figure 2.
The installed capacity of hydropower worldwide [4].

3. Hydropower as potential energy sources

Hydropower happens to be one of the major sources of electrical energy as it is clean, renewable and environment friendly as compared with fossil fuels or even the nuclear power [5]. Some of the major advantages of hydropower are:

1. Renewable source of energy with water as its primary source of energy (i.e. fuel),

2. Available capacity in abundance throughout globe with considerable potentials in developing countries,

3. Its potential avenue for flood control, contribution for networking to irrigation, aquatics farming and water parks,

4. Constant capacity and its enhancement during raining seasons.

World today hold considerable shares of electrical energy that are generated from renewable sources which is growing fast.

It is evident from the (**Figure 3**) below that hydropower hold 50% shares with 1,172 GW as compared with other renewable energy sources for electricity generation in 2018 where renewable energy accounts for the capacity of 2,351 GW. The share of hydropower in 2019 has increased to 1308 GW showing a significant increase in hydropower [6] despite the rising increases in solar and wind power generation capacities. Furthermore, IRENA statistic 2021 shows that of more than 80% install capacity which are from renewable energy in 2020 and the global renewable energy generation capacity reaches to 2799 GW in the end of same.

Though hydropower happens to be the major contributors for electricity generations, it has come into growing concern and threats. Water is the primary source of energy used which has been impacted due to climate changes. There are evidences of water bodies in rivers and streams drying up, prevailing situations of drought and glaciers meltdown. The actions to retain water bodies in this era of climate change needs to be device through collective initiatives ranging from policies, regulations, supports and mitigations measures.

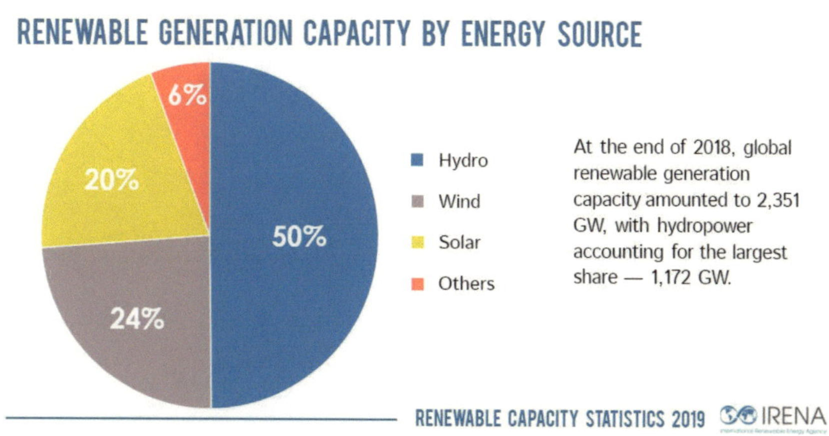

Figure 3.
The share of renewable energy generation sources [3].

4. Hydropower and sustainability

4.1 Understanding sustainable development

The concept of sustainable development has emerged in later parts of 1960s and in the earlier part of 1970s when the focuses on green movements was taking the momentum [7]. Environment concern has thus been the topic of discussion and that is to assure the sustainability (**Figure 4**).

There on the developmental activities are seen as an agent of environmental impacts and the focuses was shifted towards sustainable development. Sustainable development focuses more on the efficient measures of economic developmental activities that can be more in equitable manner and subsequently have limited impacts created to the environment (**Figure 5**).

Also, the dimensions of sustainability are well addressed in Sustainable Development Goals (SDGs) where the SDGs are measures devised to build the future that are sustainable, prosperous as well as equitable. The noble initiatives of SDGs are to provide a framework for addressing the needs of sustainable future through the initiatives and measures that works in broader aspects of activities that are planned for economic growth as well as living standards. This demand for strategic management which basically is 'understanding an organization's strategic position, making strategic decisions for the future, and managing [that] strategy in action' that can address the three phases of development as reflected below (**Figure 6**).

4.2 Understanding sustainable development of hydropower

Hydropower has to fit in the sustainability development models as discussed in earlier section where it needs to take account of economic, social and environment dimension starting from its planning phase till operations. Though these three factors are mostly in a nexus form when we are talking about sustainability.

The hydropower dams are main components of hydropower where it need to be constructed in water abundant regions. On the other hand, due to climate change the risk of drought and serviceability of the dams in meeting the needful impacts are in rise (**Figure** 7) [11].

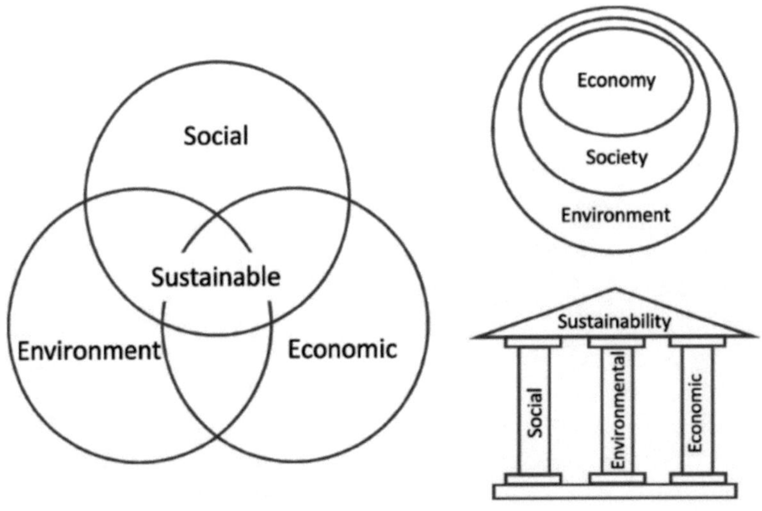

Figure 4.
Left, typical representation of sustainability as three intersecting circles. Right, alternative depictions: Literal 'pillars' and a concentric circles approach [8].

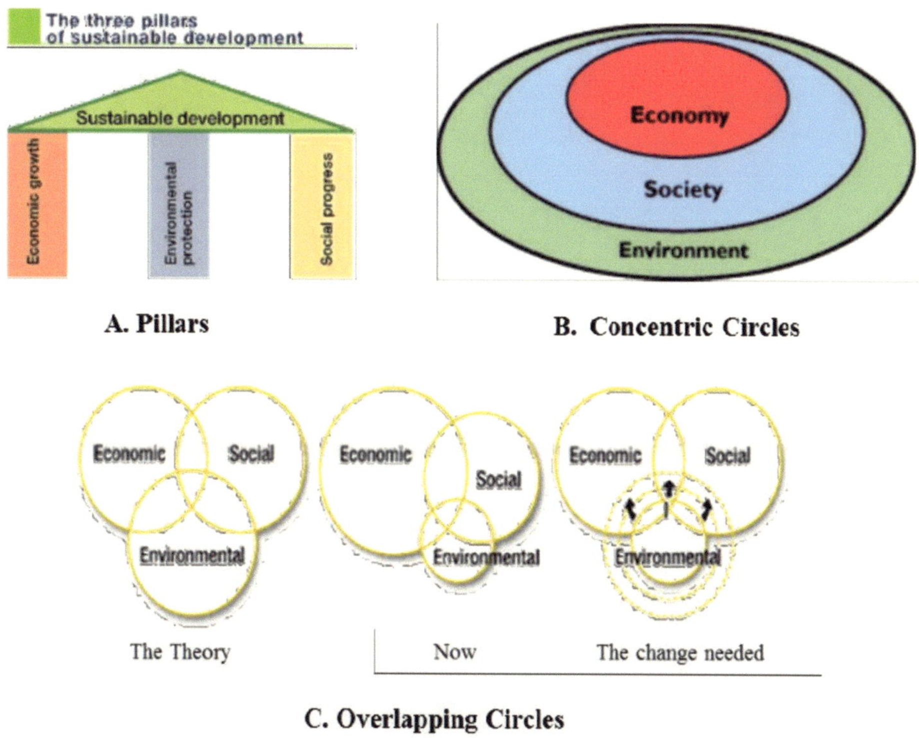

Figure 5.
Three visual representations of sustainable development: Pillars, circles, interlocking circles (IUCN, 2006) [9].

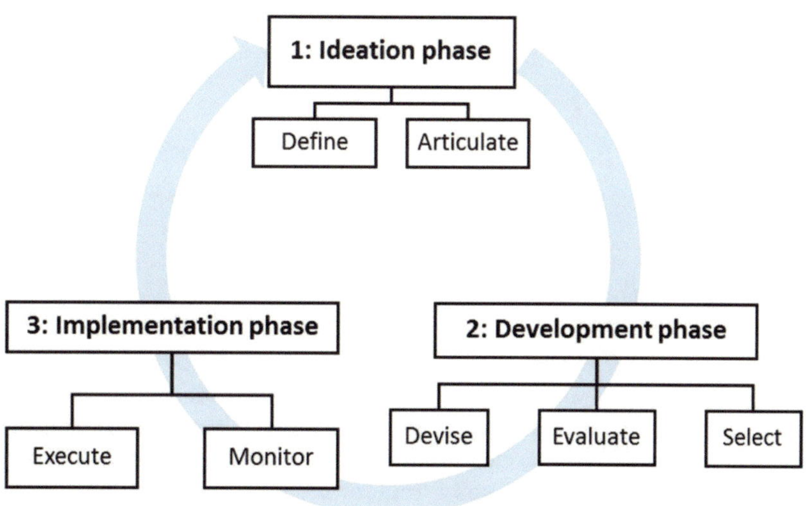

Figure 6.
A generalized strategic management process [10].

The (**Figure 7**) further clearly highlighted the rising risk of draught which also reported that around 31% of the dams that are 'planned/will be planned' will face the consequences of draught. Such threats will have serious implications on those countries which has major reliance in its power requirements from hydropower.

Other equal risk associated as a contribution of rising climate change are due to the floods that will be of risks to the dams associated with hydropower (**Figure 8**).

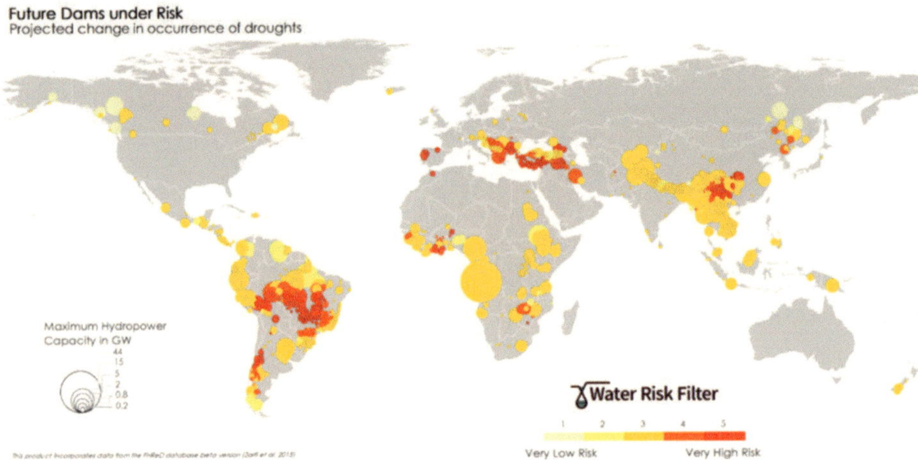

Figure 7.
The projected change in occurrence of draughts and risk to hydropower dams [11].

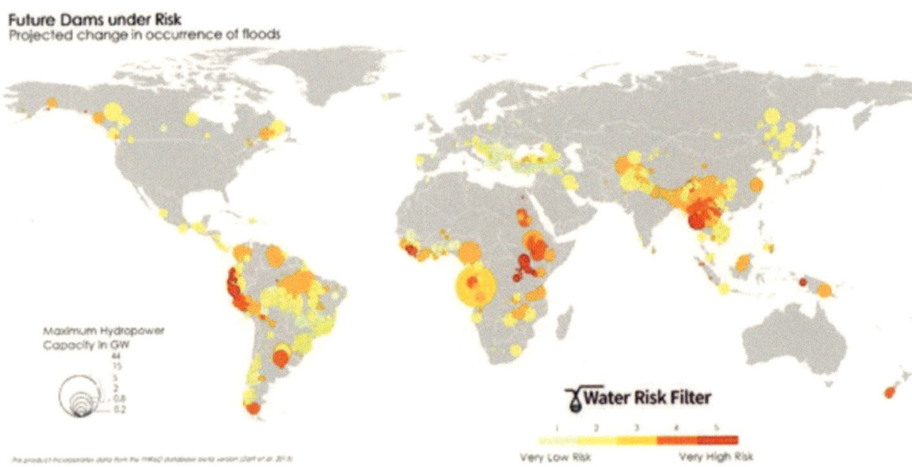

Figure 8.
The projected changes in occurrence of floods and risk to hydropower dams [11].

The figure and the resources clearly highlighted the growing risk that are poised to hydropower dams now and for future dams making it more venerable and may result into catastrophic consequences if the dams fail resulting in the devastating impacts. Though hydropower is clean source of energy, the functional loss of a hydropower reservoir's capabilities as a result of sedimentation or siltation could have both economic and environmental consequences [12].

Hydropower development is costly affairs and it can further be worsened with unforeseen complex challenges pertaining to geology as well as technical. Such situations can also stimulate the unexpected increases of the project duration in construction phase and subsequent threats during the operations demanding more attention [13].

Researchers has pointed out the needs for sustainability assessment tools. The integrated model of hydropower sustainability assessment has been also proposed where 'conceptual framework', and appropriate 'indicators selection' has been identified where the former is quite simple and practical tool and later is more of selecting indicators taking stock of environment, economic as well as social aspects into consideration. The model proposed are as shown in **Figure 9** [14]:

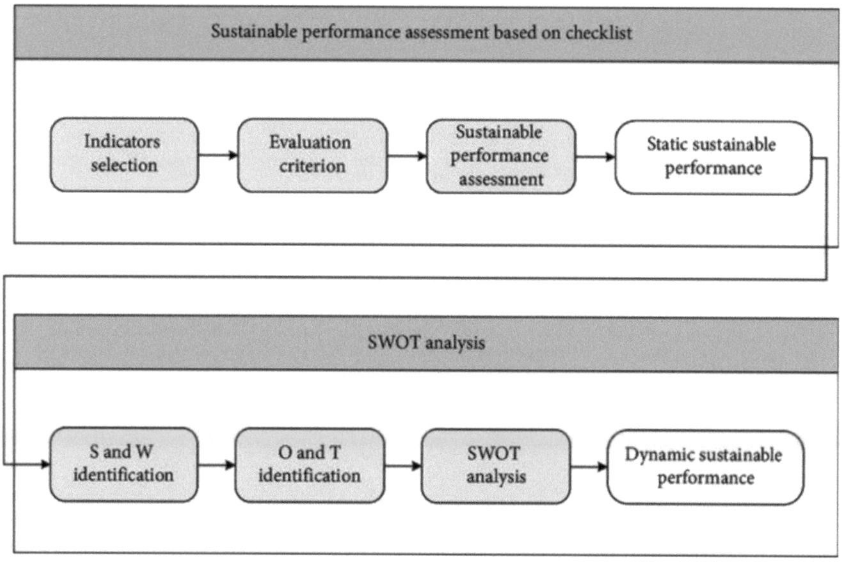

Figure 9.
Conceptual framework of the integrated hydropower sustainability assessment [14].

The above study further concluded that especially while considering the impacts associated with sustainability in case of integrated framework need to consider threats like very tight fiscal allocations and economic downturns that could further reduce sustainability.

Furthermore, there are series of risks as well as uncertainty associated with hydropower project. Some pertaining risks includes environmental, social, economic, policies and regulations, technological and financial, natural and many more which needs to be managed else threatening the availability along with sustainability of hydropower. This calls for effective risk managements in hydropower for ascertaining the sustainability [15]. On top of this one should clearly understand the internal and external risks which has direct impacts on hydropower sustainability. Internal risks are more associated with planning, accusation and execution measures including cost whereas the external factors mostly would be associated with weather, natural calamities and political measures.

Using expert judgment and a multi-criteria scoring technique, assess the potential risks of a hydro energy project. This risk assessment tool examines technical, economic, environmental, social, and regulatory threats. Researchers has presented some of the pertaining risks, its impacts and the mitigation measures as proposed are as shown in **Figure 10**.

The same figure also has shown how risks, impacts and mitigation can be looked in case of hydropower that need to be looked from the view of achieving the sustainability as well as ensuring the safe operation. Impacts level needs to be critically viewed and measures that need to be devised. Such risks, impacts and mitigation will be varied based on the location and capacity of the hydropower along with prevailing geographical as well as political issues.

This study found out that site geology and environmental issues are two external risks for hydropower projects. Hence risk assessment tools will prove handier in cost cutting as well as enhance deeper understanding of risks associated (**Figure 11**).

The appropriate planning and implementation are crucial as this will identify the actual size of hydro power plant, exact location of dam and other

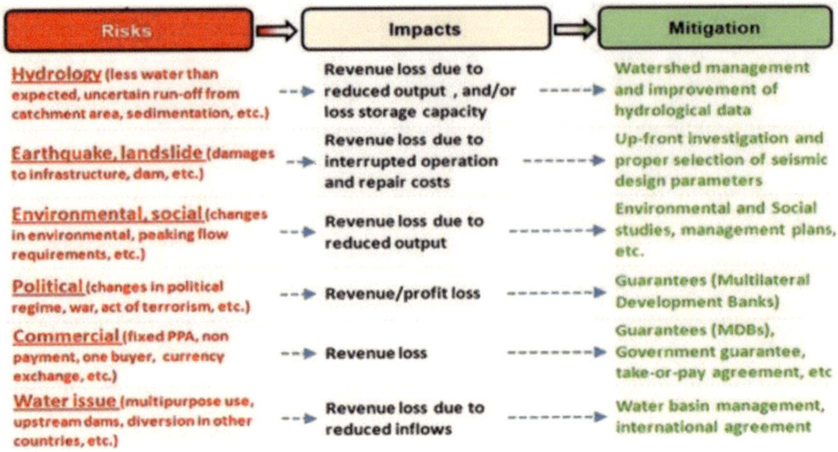

Figure 10.
The risks, impacts and mitigation approaches [16].

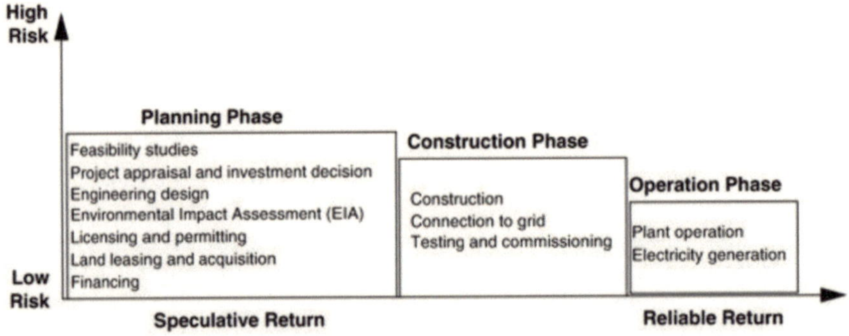

Figure 11.
*The phase wise risk assessment (**Figure 11**) [16].*

infrastructures, the right size and type of dam, right rating of turbine its capacity and types. Such system level approaches need to be materialized through proper research and evidences during the design phase whereas proper executions for quality assurance during the construction phases. Furthermore, the conditional aspects during its operation will hold equal importance in assurance sustainability issues pertaining to hydropower plant as well as power system networks as a whole.

The Low Impact Certification for the certification of hydropower with major focuses on ecological impacts by the Low Impact Hydropower Institute (LIHI) was initiated in 2000. Subsequent to it the Green Hydro Power certification was initiated by the Swiss Federal Institute of Aquatic Science and Technology (EAWAG) which is more to safeguard the ecological integrity of river system. The inconsistency in sustainability assessment of hydropower have been highlighted by researchers and the development of hydropower sustainability assessment protocol (HSAP) by International Hydropower Association (IHA) which was launched in 2008 which aims in providing the more enhanced assessment tool for hydropower sustainability. This HSPA protocol has been extensively used for the development of multiple hydropower of large-scales including the 'Three Gorges Hydropower' in China. This protocol was endorsed by World Bank in 2014 recognizing it as a tool for developing hydropower [17].

Researchers pointed out that more of methodologies and models were incorporated for sustainable hydropower development but those were more of theoretical frameworks and there is a need for quantitively evaluate the sustainability aspects of the hydropower. One method to address this as proposed by researchers are information entropy and dissipative structure theory which are combined together so as to provide an appropriate method for evaluation research in terms of sustainable hydropower development capacity [18].

There is need to understand the environmental as well as social effects of the hydropower as these two factors are most significant indicators which are highlighted in case of sustainable hydropower. Hence the use of 'Environmental and Social Impact Assessment (ESIA) and 'Strategic Environmental Assessment (SEA) are implemented too. These ESIA and SEA are two tools that are used for addressing the impacts from various sectors (i.e. infrastructure, energy and mining as major sectors and processing and manufacturing sectors as minor sources). ESIA and SEA are internationally practiced, often legally enshrined, tools for assessing the consequences of policies, plans, programs, and projects from an integrated SDG perspective [19].

The innovative solutions in context of hydropower development especially the dams and other major infrastructures needs to be explored so as to achieve hydropower sustainability as there is tremendous potentials and benefits from hydropower. In the meantime, to address sustainability the incorporation of climate change needs to be incorporated especially in building the dams associated with large scales hydropower [20].

5. Conclusions

This chapter highlighted the need of hydropower thinking from the perspective of sustainability. As hydropower projects happen to be one of the main contributors in power sector and that too a promising renewable source, there are serious questions in relation to sustainability. As the sustainability aspects of hydropower has to be address from planning, construction as well as operation stages of hydropower, it requires validated tools and protocols that have been tested and verified. This is essential so address the risk associated with hydropower from internal (those that are pertaining due to types, capacities, expected working environments etc. for hydropower which need to be taken care during planning and design stage) as well as the external factors. Usually, the external risks are major sources of impacts to sustainability of hydropower a clear insight and needful intervention where possible is needed in timely manner. The two vital external risks of concern covering major impacts are ecological and social risks factors which has to be quantitively analyzed and address so as to ascertain the sustainability of the hydropower project and making it as a promising investment option for meeting growing demand of electric power from renewable energy sources.

Author details

Hemlal Bhattarai
Centre for Lighting and Energy Efficiency Studies (CLEES), Jigme Namgyel Engineering College, Royal University of Bhutan, Dewathang, Bhutan

*Address all correspondence to: b.hemlal@gmail.com

References

[1] Chandy, T., Keenan, R. J., Petheram, R. J., & Shepherd, P. (2012). Impacts of hydropower development on rural livelihood sustainability in Sikkim, India: Community perceptions. Mountain Research and Development, *32*(2), 117-125. https://doi.org/10.1659/mrd-journal-d-11-00103.1

[2] IRENA. (2020). Renewables accounted for 72% of new power capacity globally in 2019. https://bioenergyinternational.com/heat-power/renewables-accounted-for-almost-75-of-new-power-capacity-in-2019-irena

[3] IRENA (2021). Renewable capacity statistics 2021 International Renewable Energy Agency (IRENA), Abu Dhabi

[4] IHA releases (2019). Hydropower Status Report charting growth in renewable hydro

[5] RENÖFÄLT, B., JANSSON, R., & NILSSON, C. (2010). Effects of hydropower generation and opportunities for environmental flow management in Swedish riverine ecosystems. Freshwater Biology, 55(1), 49-67. https://doi.org/10.1111/j.1365-2427.2009.02241.x

[6] IHA (2020), Hydropower Status Report. https://hydropower-assets.s3.eu-west-2.amazonaws.com/publications-docs/2020_hydropower_status_report.pdf

[7] Schumann K, Saili L, Taylor R, Abdel-Malek R. (2010). Hydropower and Sustainable Development: A Journey. World Energy Congress 2010. International Hydropower Association

[8] Purvis, B., Mao, Y., & Robinson, D. (2018). Three pillars of sustainability: In search of conceptual origins. Sustainability Science, 14(3), 681-695. https://doi.org/10.1007/s11625-018-0627-5

[9] Jalaei, F., & Jrade, A. (2015). Integrating building information modeling (BIM) and LEED system at the conceptual design stage of sustainable buildings. Sustainable Cities and Society, 18, 95-107. https://doi.org/10.1016/j.scs.2015.06.007

[10] Grainger-Brown, J., & Malekpour, S. (2019). Implementing the sustainable development goals: A review of strategic tools and frameworks available to organisations. Sustainability, 11(5), 1381. https://doi.org/10.3390/su11051381

[11] WWF. (2019, December 13). The power of rivers at risk: Mapping rising threats to hydropower dams using WWF's water risk.... Medium. https://wwf.medium.com/the-power-of-rivers-at-risk-mapping-rising-threats-to-hydropower-dams-using-wwfs-water-risk-eb13bb041082

[12] Luis, J., Sidek, L. M., Desa, M. N., & Julien, P. Y. (2013). Sustainability of hydropower as source of renewable and clean energy. IOP Conference Series: Earth and Environmental Science, 16, 012050. https://doi.org/10.1088/1755-1315/16/1/012050

[13] World Bank. (2020, May 26). Project Information Document (PID). https://documents1.worldbank.org/curated/en/930481591860622408/text/Concept-Project-Information-Document-PID-Agus-Pulangi-Hydropower-Complex-Rehabilitation-Project-P173728.tx

[14] Xu, Z., Niu, Y., Liang, Y., Li, Z., & Iftikhor, A. (2020). The integrated hydropower sustainability assessment in Tajikistan: A case study of Rogun hydropower plant. Advances in Civil Engineering, *2020*, 1-18. https://doi.org/10.1155/2020/8894072

[15] Shaktawat, A., & Vadhera, S. (2020). Risk management of

hydropower projects for sustainable development: A review. Environment, Development and Sustainability, 23(1), 45-76. https://doi.org/10.1007/s10668-020-00607-2

[16] Kucukali, S. (2014). Risk assessment in Hydroenergy projects: Learning from experts and data. Handbook of Renewable Energy, 1-17. https://doi.org/10.1007/978-3-642-39487-4_5-1

[17] Tahseen, S., & Karney, B. W. (2017). Reviewing and critiquing published approaches to the sustainability assessment of hydropower. Renewable and Sustainable Energy Reviews, 67, 225-234. https://doi.org/10.1016/j.rser.2016.09.031

[18] Liang, X., Si, D., & Xu, J. (2018). Quantitative evaluation of the sustainable development capacity of hydropower in China based on information entropy. Sustainability, 10(2), 529. https://doi.org/10.3390/su10020529

[19] Netherlands Commission for Environmental Assessment. (2017). The role of ESIA and SEA in mainstreaming biodiversity (ESIA and SEA for Sustainable Hydropower Development). Convention on Biological Diversity. https://www.cbd.int/mainstreaming/doc/seas-hydropower-en.pdf

[20] Moran, E. F., Lopez, M. C., Moore, N., Müller, N., & Hyndman, D. W. (2018). Sustainable hydropower in the 21st century. Proceedings of the National Academy of Sciences, 115(47), 11891-11898. https://doi.org/10.1073/pnas.1809426115